CHEMISTRY RESEARCH AND APPLICATIONS

RARE EARTH ELEMENTS: SUPPLY CHAIN RISKS AND NATIONAL DEFENSE ISSUES

CHEMISTRY RESEARCH AND APPLICATIONS

Additional books in this series can be found on Nova's website under the Series tab.

Additional e-books in this series can be found on Nova's website under the e-book tab.

CHEMISTRY RESEARCH AND APPLICATIONS

RARE EARTH ELEMENTS: SUPPLY CHAIN RISKS AND NATIONAL DEFENSE ISSUES

KRISTIN WATERS
EDITOR

NOVINKA

New York

Library of Congress Cataloging-in-Publication Data

ISBN: 978-1-53610-120-1

Published by Nova Science Publishers, Inc. † New York

CONTENTS

PREFACE

The Department of Defense (DOD) depends on rare earths that contain one or more of 17 similar metals which have unique properties, such as magnetism at high temperatures, to provide functionality in weapon system components. Many steps in the rare earths supply chain, such as mining, are conducted in China, a situation that may pose risks to the continued availability of these materials. The Joint Explanatory Statement accompanying the Carl Levin and Howard P. "Buck" McKeon National Defense Authorization Act for 2015 included a provision for the United States Government Accountability Office (GAO) to review DOD efforts to identify and mitigate risks in its rare earths supply chain. This book assesses the extent that DOD determined which rare earths, if any, are critical to national security; and has identified and mitigated risks associated with rare earths, including the effects of a potential supply disruption.

In: Rare Earth Elements
Editor: Kristin Waters

ISBN: 978-1-53610-120-1

Chapter 1

RARE EARTH MATERIALS: DEVELOPING A COMPREHENSIVE APPROACH COULD HELP DOD BETTER MANAGE NATIONAL SECURITY RISKS IN THE SUPPLY CHAIN*

United States Government Accountability Office

WHY GAO DID THIS STUDY

DOD depends on rare earths that contain one or more of 17 similar metals which have unique properties, such as magnetism at high temperatures, to provide functionality in weapon system components. Many steps in the rare earths supply chain, such as mining, are conducted in China, a situation that may pose risks to the continued availability of these materials. The Joint Explanatory Statement accompanying the Carl Levin and Howard P. "Buck" McKeon National Defense Authorization Act for 2015 included a provision for GAO to review DOD efforts to identify and mitigate risks in its rare earths supply chain.

This report assesses the extent that DOD (1) determined which rare earths, if any, are critical to national security; and (2) has identified and mitigated

* This is an edited, reformatted and augmented version of a United States Government Accountability Office, Publication No. GAO-16-161, dated February 2016.

risks associated with rare earths, including the effects of a potential supply disruption. GAO reviewed DOD reports from 2011-2015 and relevant legislation; and collected information from DOD, the military departments, and industry organizations.

What GAO Recommends

GAO recommends that DOD designate which rare earths are critical to national security, and develop a comprehensive approach to help ensure a secure supply by identifying risk metrics, among other activities. DOD concurred with all the recommendations in this report and provided timeframes for action.

What GAO Found

Three Department of Defense (DOD) offices have identified certain rare earth materials (rare earths) as critical for some defense applications, such as lasers, but DOD has not taken a comprehensive, department-wide approach to identifying which rare earths, if any, are critical to national security. Specifically, DOD offices have not yet agreed on what constitutes "critical" rare earths. Using different statutorily-based definitions, these offices have identified 15 of the 17 rare earths as critical over the last 5 years (see table).

DOD's current approach to identifying and mitigating risks associated with rare earths is fragmented. With different interpretations of which rare earths are critical, establishing priorities to analyze supply risk becomes difficult. For example, the Defense Logistics Agency-Strategic Materials office methodically analyzes risks for all materials, but its focus is a four-year timeframe with stockpiling as its mitigation tool. The Manufacturing and Industrial Base Policy office relies on other DOD organizations to identify and elevate risks, relies primarily on the market to resolve supply disruptions, and has not put in place measures to evaluate the success of its mitigating actions. According to DOD, supply disruptions in rare earths have not occurred over the last several years. Regardless, the Strategic Materials Protection Board has not developed a comprehensive approach for ensuring a sufficient supply of rare earths for national security needs—one that can establish criticality, assess supply risks, and identify mitigating actions. Such an approach would better position DOD to help ensure continued functionality in weapon system components should a disruption occur.

Rare Earth Materials (Rare Earths) Meeting Various Definitions of Critical by Office

Rare earths	Acquisition, Technology and Logistics 2012	Manufacturing and Industrial Base Policy (MIBP)[a] 2013	Defense Logistics Agency-Strategic Materials 2011, 2013, 2015
Scandium			•
Yttrium	•	•	•
Lanthanum			•
Praseodymium	•		•
Neodymium	•		•
Samarium			•
Europium	•		•
Gadolinium	•		
Terbium		•	•
Dysprosium	•	•	•
Holmium		•	
Erbium	•	•	•
Thulium		•	•
Ytterbium		•	
Lutetium		•	

Source: GAO presentation of Department of Defense reports. I GAO-16-161

Notes: Two rare earths, cerium and promethium, were not identified as critical. The rare earths are an International Union of Pure and Applied Chemistry named group.

[a]MIBP identified these rare earths as higher risk.

ABBREVIATIONS

DLA-Strategic Materials	Defense Logistics Agency-Strategic Materials
DOD	Department of Defense

MIBP	Office of the Deputy Assistant Secretary of Defense for Manufacturing and Industrial Base Policy
NDAA	National Defense Authorization Act
OUSD(AT&L)	Office of the Undersecretary of Defense for Acquisition, Technology and Logistics
Rare earths	Rare earth materials
SMPB	Strategic Materials Protection Board

* * *

February 11, 2016

Congressional Committees

The Department of Defense (DOD) depends on rare earth materials (rare earths) to provide functionality in weapon systems components.[1] Many steps in the rare earths supply chain, such as mining and refining the ore, are primarily conducted outside the United States, which may pose risks to continued availability of these materials to DOD. For example, in our prior work conducted in 2010, we found that much of the rare earths processing is performed in China, giving it a dominant position that could affect worldwide supply and prices.[2] U.S. industry previously performed all steps in the rare earths supply chain and produced the majority of the global supply but no longer has the capability. DOD reported in 2014 that there have been no availability issues and expects none in the near term. DOD has been generally directed by law since at least 2011 to take actions concerning supply chain vulnerabilities for materials, such as rare earths. The Defense Logistics Agency-Strategic Materials (DLA-Strategic Materials) has had management responsibility since 1988 to store select materials—which may include rare earths—in the National Defense Stockpile to mitigate potential shortages.

The Joint Explanatory Statement accompanying the Carl Levin and Howard P. "Buck" McKeon National Defense Authorization Act (NDAA) for Fiscal Year 2015 included a provision for GAO to review DOD efforts to manage risks in its rare earths supply chain, including DLA-Strategic Materials' stockpiling efforts. We assessed the extent that DOD (1) determined which rare earths, if any, are critical to national security; and (2) has identified and mitigated risks associated with rare earths, including the effects of a potential supply disruption.

To determine the extent DOD identified which rare earths, if any, were critical to national security, we reviewed DOD reports from 2011 to 2015 to understand the analyses and criteria that DOD conducted and used to determine critical rare earths. We also reviewed the actions of DOD components—Air Force, Army, and Navy—to determine if they had identified critical rare earths. In addition, we interviewed DOD and other agencies' officials, such as at the United States Geological Survey, with defense-related industrial base knowledge.

To assess the extent to which DOD has identified risks and taken action to mitigate the risks associated with rare earths, including the effects of a potential supply disruption, we reviewed DOD reports from 2011 to 2015 to understand the analyses conducted, risks identified, and mitigating actions. We also reviewed the actions of DOD components to determine if they had identified any risks in the rare earths supply chain. In addition, we interviewed DOD and other agencies' officials with knowledge of the defense industrial base and management of the National Defense Stockpile.

We conducted interviews and collected information for a non-generalizable sample of 20 cases selected for illustrative purposes based on rare earths and risk management processes, including three DOD offices with purview over weapon systems, three defense contractors, and seven companies. We reviewed DOD's sources of data for its analytical processes. We determined that the information that we collected were sufficiently reliable for our report. We assessed DOD's policies, procedures, and practices against criteria in applicable statutes, *Standards for Internal Control in the Federal Government*, and GAO's framework for risk assessment.[3] A more detailed discussion of our scope and methodology is provided in appendix I.

We conducted this performance audit from February 2015 to February 2016 in accordance with generally accepted government auditing standards. Those standards require that we plan and perform the audit to obtain sufficient, appropriate evidence to provide a reasonable basis for our findings and conclusions based on our audit objectives. We believe that the evidence obtained provides a reasonable basis for our findings and conclusions based on our audit objectives.

BACKGROUND

Rare earths contain one or more of the following 17 elements in the periodic table: the lanthanides that begin with lanthanum (atomic number 57) through lutetium (atomic number 71) and two non-lanthanides that have similar properties, yttrium and scandium (see figure 1).

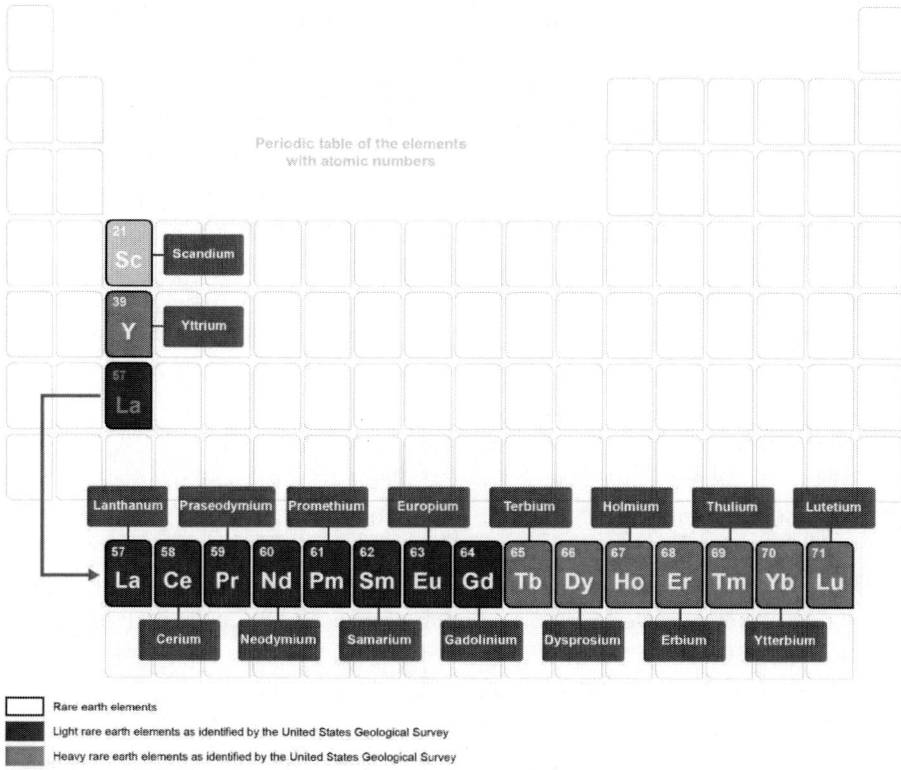

Source: GAO presentation of the rare earth elements in the periodic table. | GAO-16-161.

Figure 1. Rare Earth Elements in the Periodic Table.

These elements are referred to as rare because they appear in low concentrations in the ground—though relatively abundant overall—and are difficult and costly to mine and process. Rare earth elements are often classified as either heavy or light based, in part, on their chemical properties. According to the United States Geological Survey, heavy rare earth elements are generally less abundant and more expensive due to their scarcity, more unique characteristics, and strong demand relative to the light rare earth elements.[4] DOD has a particular interest in two heavy rare earth elements, yttrium and dysprosium.

Rare earth elements are generally found in mined ore, separated and converted into intermediary forms, such as oxides, and further processing depends on the application. For example, the intermediary form can be turned into a metal, and added to other materials to form an alloy for use in semi-finished metal components that are incorporated into end products. The United States was the leader in global production of rare earth oxides, metals, and alloys from the 1960s through the 1980s but currently has limited capability. According to a 2013 Congressional Research Service report, global production is prevalent outside the United States due, in part, to lower labor costs and more lenient environmental standards.[5] At present, the United States generally relies on imports from global manufacturers of rare earths, such as metals, alloys, and magnets. See figure 2 for a representative rare earth metals supply chain that is generally outside the United States.

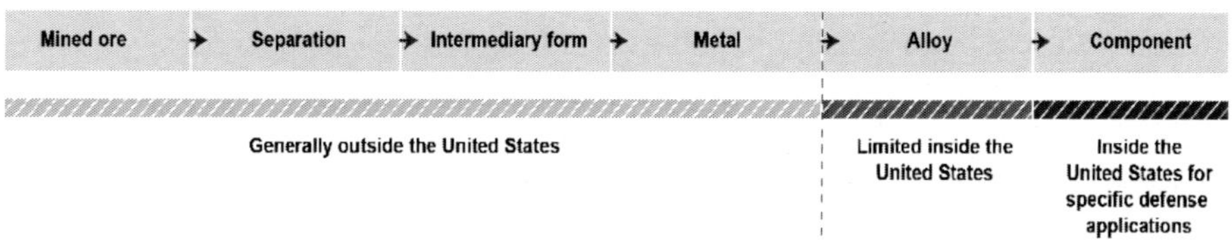

Source: GAO analysis and presentation of U.S. government and industry reports and interviews. | GAO-16-161.

Figure 2. Generalized Rare Earth Materials Supply Chain for Metal Components.

The United States is a major consumer of defense and commercial end products containing rare earths, however, its demand for rare earths is approximately nine percent of the global demand, according to DOD estimates. Also, DOD demand is approximately only one percent of the United States demand. Recent studies have shown that rare earths are essential to the production, sustainment, and operation of U.S. military equipment. Reliable access to the necessary material, regardless of the overall level of defense demand, is a bedrock requirement for DOD.[6] Some defense applications with rare earths are coatings for jet engines, missile guidance systems, antimissile defense systems, satellites, and communication systems. Examples of commercial products with rare earths are automotive catalytic converters, petroleum refining chemicals, flat panel displays, permanent magnets, and rechargeable batteries for hybrid vehicles.

Since 2010, DOD has conducted studies and published reports on various rare earths issues in response to congressional mandates (see table 1 below).

Table 1. Select Department of Defense Reports on Rare Earth Materials

Date	Report title
March 2012	*Report to Congress Rare Earth Materials in Defense Applications*
September 2012	*Report to Congress on Assessment of Feasibility and Advisability of Establishment of Rare Earth Material Inventory*
September 2012	*Report on Feasibility and Desirability of Recycling, Recovery, and Reprocessing Rare Earth Elements*
April 2013	*Recovery of Rare Earth Elements from Fluorescent Lighting Materials*
February 2014	*Diversification of Supply Chain and Reclamation Activities Related to Rare Earths*
February 2014	*Report to Congress on Implementation of Rare Earth Elements Strategy in the Joint Strike Fighter Program*

Source: GAO presentation of Department of Defense reports. | GAO-16-161.

DOD also periodically issues two reports on the defense industrial base and on "critical" and "strategic and critical" materials.[7] The *Annual Industrial Capabilities Report to Congress* provides analyses of sectors of the defense industrial base such as aircraft and ground vehicles and summarizes the department's efforts to identify the availability of materials, such as rare earths. The biennial *Strategic and Critical Materials Report on Stockpile Requirements* summarizes DLA-Strategic Materials' analyses of materials for the National Defense Stockpile.

In addition, in July 2014 the DOD Inspector General issued a report assessing whether DOD effectively planned for life-cycle sustainment of rare earth elements for the defense industrial base.[8] The DOD Inspector General found that DLA-Strategic Materials lacked procedures to validate supply and demand data for its stockpiling model. Further, the Inspector General identified internal control weaknesses related to the assessment of rare earth element supply and demand for defense applications. Specifically, the Inspector General found that DOD lacked a comprehensive and reliable approach to assess rare earth element supply and demand. The DOD Inspector General recommended that DLA-Strategic Materials have a plan for using verified and validated data to improve forecasts for the stockpile. DLA-Strategic Materials generally agreed, noting that its model is being reviewed and will be accredited.

Three organizations in the department have related statutory requirements to manage risks from DOD's use of "critical" and "strategic and critical" materials, such as rare earths. The three organizations are DLA-Strategic Materials and the Office of the Deputy Assistant Secretary of Defense for Manufacturing and Industrial Base Policy (MIBP), both in the Office of the

Undersecretary of Defense for Acquisition, Technology and Logistics (OUSD(AT&L)), and the Strategic Materials Protection Board (SMPB). Their select responsibilities are shown in table 2 below.

Table 2. Three Department of Defense (DOD) Organizations and Their Select Statutory Responsibilities for Managing Material Risks

Organization	Select responsibilities	Legal source of responsibility
Defense Logistics Agency-Strategic Materials (DLA-Strategic Materials)	• Determine materials deemed strategic and critical to national security. • Preclude, when possible, a dangerous and costly dependence by the United States upon foreign sources for supplies of such materials in times of national emergency. • Issue a biennial report to Congress detailing stockpiling recommendations for the National Defense Stockpile based on certain national emergency planning assumptions. The National Defense Stockpile maintains a domestically held inventory of strategic and critical materials, and certain other materials acquired for or transferred to the stockpile.	United States Code, Title50, Section 98, et seq., as added, amended, and delegated
Office of the Deputy Assistant Secretary of Defense for Manufacturing and Industrial Base Policy (MIBP)[a]	• Ensure reliable sources of materials critical to national security, such as specialty metals, armor plate, and rare earth elements. • Establish DOD policies for developing and maintaining the defense industrial base of the United States and for continued reliable resource availability from secure sources. • Provide policy and oversight of matters related to materials critical to national security to ensure a secure supply of such materials to DOD. • Provide recommendations and acquisition policy guidance on supply chain management and supply chain vulnerability throughout the entire supply chain, from suppliers of raw materials to producers of major end items.	United States Code, Title10, Section 139c, as amended
Strategic Materials Protection Board (SMPB) (MIBP, chair; DLA-Strategic Materials, vice-chair)[b]	• Analyze the risk associated with each material SMPB designated as critical to national security and the effect on national defense that the non availability of such material would have.	United States Code, Title10, Section 187, as amended

Table 2. (Continued)

Organization	Select responsibilities	Legal source of responsibility
	• Determine the need to provide a long term secure supply of materials designated as critical to national security to ensure that national defense needs are met. • Recommend a strategy to the Secretary of Defense to ensure a secure supply of materials designated as critical to national security and recommend other strategies as appropriate to strengthen the industrial base with respect to materials critical to national security. • Meet and publish recommendations regarding materials critical to national security not less frequently than once every 2 years in the Federal Register.	

Source: GAO summary of select DOD responsibilities from United States Code. | GAO-16-161

[a]The Deputy Assistant Secretary for Defense for MIBP is the principal advisor to the Under Secretary of Defense for Acquisition, Technology and Logistics in the performance of the Under Secretary's duties relating to the responsibilities listed here. 10 U.S.C. § 139c(b).

[b]The National Defense Authorization Act for Fiscal Year 2013 named the Deputy Assistant Secretary of Defense for MIBP as the chairman and the Administrator of the DLA-Strategic Materials, or any successor organization, as the vice-chairman.

In addition to MIBP and DLA-Strategic Materials representation, the SMPB membership is department-wide and includes designees of the Assistant Secretary of the Army for Acquisition, Logistics and Technology and the Assistant Secretary of the Navy for Research, Development, and Acquisition.

DOD Has Identified Certain Materials as Critical but Approach Not Comprehensive

From 2011 to 2015, DOD identified certain materials, including rare earths, as critical to meet its individual statutory responsibilities, but has not taken a comprehensive approach to identify which of these materials are critical to national security. Based on GAO's prior work, the Office of the Deputy Under Secretary of Defense for Industrial Policy, the predecessor of

MIBP, stated that it was collaborating with the Defense Logistics Agency's National Defense Stockpile Center to create department-wide criteria for the terms "critical" and "strategic," and expected to report the results of this effort at the end of calendar year 2008.[9] As of November 2015, there were no department-wide criteria for these terms, but DOD officials said that they were considering discussing the definitions and possibly proposing a group of materials that meet the SMPB's definition of critical to national security. SMPB and OUSD(AT&L) and two of its offices—MIBP and DLA-Strategic Materials—have separate, existing statutory definitions for critical materials. See table 3 for the varying definitions of critical materials by organization.[10]

Table 3. Critical Materials Definitions by Organization

Department of Defense Organization	Office of the Under Secretary of Defense for Acquisition, Technology and Logistics	Strategic Materials Protection Board	Manufacturing and Industrial Base Policy	Defense Logistics Agency-Strategic Materials
Terms	Critical rare earth materials in defense applications	Materials critical to national security	Materials critical to national security	Strategic and critical materials
Statutory definitions	(a) The rare earth material is critical to the production, sustainment, or operation of significant U.S. military equipment and	(a) Upon which the production or sustainment of military equipment is dependent and	(a) Upon which the production or sustainment of military equipment is dependent and	(a) Would be needed to supply the military, industrial, and essential civilian needs of the United States during a national emergency and
	(b) The rare earth material is subject to interruption of supply, based on actions or events outside the control of the government of the United States	(b) The supply of which could be restricted by actions or events outside the control of the government of the United States	(b) The supply of which could be restricted by actions or events outside the control of the government of the United States	(b) Are not found or produced in the United States in sufficient quantities to meet such need

Source: GAO presentation of statutory definitions. | GAO-16-161

Table 4. Rare Earth Materials Meeting Various Definitions of Critical by Office and Report from 2011 to 2015

Rare earth materials	Office of the Undersecretary of Defense for Acquisition, Technology and Logistics report	Manufacturing and Industrial Base Policy report	Defense Logistics Agency—Strategic Materials report on stockpile requirements	Defense Logistics Agency—Strategic Materials report on stockpile requirements	Defense Logistics Agency—Strategic Materials report on stockpile requirements
	2012	2013	2011	2013	2015
Scandium				●	
Yttrium	●	○[a]	●	●	●[b]
Lanthanum					●
Cerium					
Praseodymium	●		●		
Neodymium	●		●		
Promethium					
Samarium			●		
Europium	●		●		●
Gadolinium	●				
Terbium		○	●	●	
Dysprosium	●	○	●	●	
Holmium		○			
Erbium	●	○		●	
Thulium		○		●	
Ytterbium		○			
Lutetium		○			

Source: GAO presentation of Department of Defense Reports. | GAO-16-161.

[a]Manufacturing and Industrial Base Policy identified heavy rare earth materials as a category at a higher risk for a supply disruption in 2013. It has not identified any rare earth materials as meeting its definition of materials critical to national security.

[b]Defense Logistics Agency–Strategic Materials specifically identified high purity yttrium oxide.

The DOD organizations, generally using their respective definitions and analyses for their individual responsibilities, identified between three and eight of the rare earths as having met the definitions at least once from 2011 to 2015. Overall, 15 of the 17 rare earths were identified as critical at some point in the last 5 years. None of the organizations identified cerium or promethium as having met the definitions during that time period. See table 4 for the rare earths that each DOD organization separately identified as critical in defense applications, critical to national security, or as strategic and critical in different reports.

DOD's identification of which rare earths, if any, are critical to national security is fragmented.[11] Three DOD organizations separately identified five different lists of rare earths during this time frame using their respective definitions of critical to national security. According to GAO's risk management framework, comprehensive management of risk is a necessary step for agencies, like DOD, to manage and assess risk to help make informed decisions and prioritize resource investments.[12] The SMPB, with the duty of designating specific materials as critical to national security, has reported to Congress in 2007, 2008, and 2011 that its focus is to assess the criticality of materials and to identify which materials are critical to national security, but has not done so.[13] The SMPB acknowledged that identifying the materials critical to national security is necessary to ensure consistency among the DOD offices. However, in a February 2014 letter from MIBP to DLA-Strategic Materials, the respective offices of the chair and vice-chair of the SMPB, that discusses cooperation concerning the SMPB, the designation of materials critical to national security was not one of the areas outlined for proposed coordination. DOD officials told us that they are generally working toward coordinating MIBP's sector analyses of the defense industrial base—such as aircraft and ground vehicles—with DLA-Strategic Materials' analyses that support recommendations for materials to stockpile in order to have a more comprehensive understanding of supply chains.

DLA-Strategic Materials' responsibility for determining materials to stockpile under certain scenarios has been considered by some within DOD as a means to ensure materials critical to national security are available. For example, MIBP officials told us that they rely on DLAStrategic Materials' analyses and expertise for other responsibilities such as MIBP's Title 10 responsibilities of ensuring reliable sources of materials critical to national security. However, DLA-Strategic Materials officials caveat their analyses as being valid for a single point in time and for decisions related to their title 50 responsibilities for the National Defense Stockpile, and not for other statutory responsibilities such as identification of materials critical to national security in the Title 10 definitions. To demonstrate that its analyses are valid for a specific point in time, DLA-Strategic Materials officials indicated that the identification of rare earths that meet the definition for strategic and critical may differ over time, unlike OUSD(AT&L)'s and MIBP's lists, which have not changed. In addition, the DOD-created emergency scenarios may change for each biennial analysis, which affects DLA-Strategic Materials identification of which materials meet the definition for strategic and critical. Moreover, DLA-Strategic Materials officials told us that they are not responsible for designating if a material is strategic and critical but are tasked with determining if a material is estimated to have a shortfall in a given year based upon DOD emergency scenarios.

In addition to DOD's differing lists of critical rare earths, weapon system officials at the military departments that we spoke with also have identified what they consider as critical rare earths in their systems, but this information is not reported consistently or department-wide. These officials do not have a definition for critical materials nor do they have an agreed upon department-wide list of critical rare earths, but they know which rare earths provide enabling functionality in the weapon systems. For example, a subject matter expert on lasers at Naval Surface Warfare– Crane told us that the rare earth component in certain defense lasers is what creates and focuses the light beam. In another example, officials at the Army Program Executive Office for Ammunition told us that rare earth magnets enable guided artillery ammunition to move in flight to the target. Further, Army Research Laboratory officials told us that they could better focus their work on specific rare earths if there was a department-wide list of rare earths critical to national security.

Implementation of controls consistent with federal internal control standards provides management with some added confidence regarding the achievement of objectives, provides feedback on how effectively an entity is operating, and helps reduce risks affecting the achievement of the entity's objectives. In particular, these standards state that management is to use data to create quality information that is appropriate, current, complete, accurate, accessible, and provided on a timely basis; and internally communicate the necessary quality information. Information and communication are necessary to create an effective internal control system that informs the allocation of resources in relation to the areas of greatest risk relevant to achieving the entity's objectives, such as efficient operations.[14]

From an industry perspective, we spoke with five defense and petroleum corporations that took steps to assess risk to help make informed decisions and prioritize resource investments. Company officials we spoke with generally said that they determine what materials are critical by obtaining data and communicating across product lines. Officials from three corporations stated that they have a process to gather data throughout the company and decide what materials may be critical to the company as a whole. For example, an official from one company stated that it regularly collects data from manufacturing divisions across the company and differentiates critical materials by their ability to create a production line stoppage. For potential material shortages that can stop production, for example, company officials may have regular meetings with the suppliers, forecast supplies 6 months into the future, adjust contracting arrangements, and have mitigation actions in place.

Given the differing interpretations of what rare earths are critical according to the individual responsibilities of each organization, DOD has not taken the next step to identify an appropriate, complete, timely list of department-wide rare earths critical to national security. Without taking a comprehensive department-wide approach to determine which rare earths, if any, are critical to national security, DOD is not in the best position to communicate and effectively focus resources across the department and align efforts to ensure the supply of rare earths critical to national security.

DOD Identifies Supply Disruption Risk in a Limited Manner and Has Not Proactively Analyzed the Impact of Rare Earths Unavailability

Rare earths risk assessment is, in part, a function of the probability of a supply disruption and the vulnerability of a weapon system or supply chain to the disruption. Although DOD regularly identifies supply disruption risks, it has not taken a comprehensive approach to assess and mitigate those risks, including analyzing the effect of unavailability of rare earths designated as critical to national security which is statutorily required. Fortunately, these types of supply disruptions have not materialized over the last several years, but DOD's current ability to respond may not be sufficient should they materialize in the future. Specifically, DLA-Strategic Materials estimates the risk of supply shortfalls through a biennial assessment process and makes recommendations for acquiring certain materials for the National Defense Stockpile. MIBP also has taken some action to identify supply disruption risks, but its mitigating strategy is reactive in nature. However, the SMPB, chaired and vice-chaired by officials of MIBP and DLA-Strategic Materials respectively, with the responsibility for analyzing the effect of unavailability of materials such as rare earths critical to national security may have on national defense, has not done so. Moreover, DOD has not established internal controls to measure risk and the effectiveness of its mitigating actions. Rather, DOD has taken actions without knowing the extent of the underlying risks that the unavailability of rare earths would have on its weapon systems specifically and national defense generally.

DLA-Strategic Materials Estimates the Risk of Supply Shortfalls Using National Emergency Scenarios and Recommends Materials to Stockpile

DOD identifies potential supply disruption risks and takes some mitigating actions through DLA-Strategic Materials' biennial process of assessing material shortfall risks and making recommendations to acquire certain materials for the National Defense Stockpile. The assessments are not designed to be comprehensive but are limited to a four-year time frame related to DOD defined national emergencies. The resulting shortfall risks are an estimated deficiency of materials for essential civilian and defense needs during a period of recovery after accounting for existing mitigating options, such as substitution of other materials to fill the deficiency. In each of its 2011, 2013, and 2015 assessments, DLA-Strategic Materials considered between three and seven specific rare earths at risk for a shortfall—15 in all.

The assessment process is limited to DOD-created national emergency scenarios. These scenarios cover a four-year time frame in which the first year is assumed to be a period of conflict, and years two through four are assumed to be a period of recovery. The one year period of conflict takes into consideration the following conditions: (1) a catastrophic attack on a U.S. city by a foreign terrorist organization or rogue state, (2) two near simultaneous major combat operations, (3) conflict damage from a highly capable enemy, and (4) ongoing military operations such as a military presence in a foreign country. According to DOD officials, rare earths from the National Defense Stockpile have not been used.

DLA-Strategic Materials' process to assess material shortfalls, in order to meet select statutory requirements, is methodical and has many steps. Among the ongoing data collection and analysis efforts, first, DLAStrategic Materials includes materials of concern on a "watch list" based on input from other DOD organizations. Second, it conducts initial research to determine which materials on the watch list to further analyze based on the potential for a supply disruption of the materials that may lead to a shortfall deficiency. This initial research is, in part, based on demand and supply data from industry and public sources. Third, DLAStrategic Materials uses estimating models to assess shortfall risks. Fourth, to address mitigating actions, DLA-Strategic Materials uses its judgment to determine which materials with shortfall risks that it will recommend for stockpiling to the Congress. See figure 3 for materials that DLA-Strategic Materials assessed in 2015, including rare earths.[15]

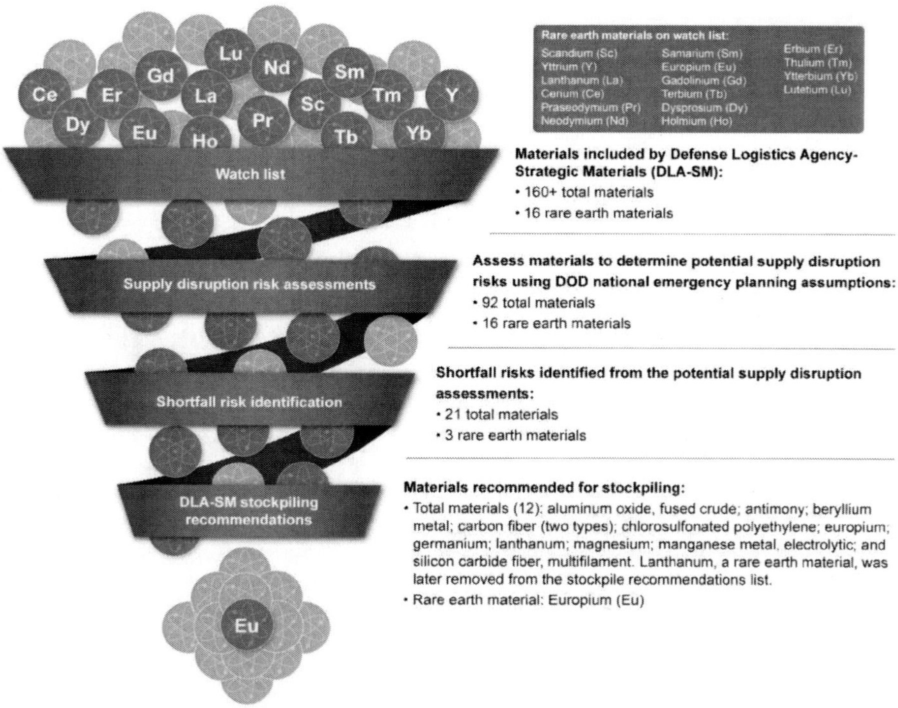

Source: GAO analysis of Defense Logistics Agency-Strategic Materials report from 2015. | GAO-16-161.

Figure 3. Defense Logistics Agency-Strategic Materials 2015 Assessment of Materials and Stockpiling Recommendations.

DLA-Strategic Materials may conduct additional research, referred to as "deep dives," on select materials or products to inform its decision on materials to recommend for stockpiling. See table 5 for a list of additional research conducted on rare earths from 2011 through 2015.

Table 5. Defense Logistics Agency-Strategic Materials Deep Dives on Rare Earth Materials from 2011 to 2015

Rare earth material	Time frame
Terbium	2011
Neodymium iron boron magnets	2011-present
Samarium cobalt magnets	2011-2012
Yttrium	2012-2015
Scandium	2013-2015
Recycled rare earth materials	2013-2014
Europium	2015 (pending)
Ytterbium	2015 (pending)

Source: GAO analysis of Defense Logistics Agency-Strategic Materials reports from 2011 to 2015. | GAO-16-161

To mitigate shortfall risks, DLA-Strategic Materials recommended stockpiling dysprosium and yttrium in 2013, and europium in 2015. See figure 4 below for the rare earths DLA-Strategic Materials identified with shortfall risks and recommended for stockpiling.

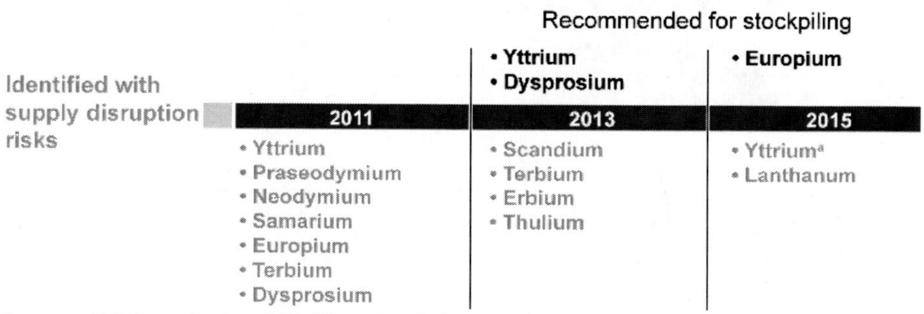

Source: GAO analysis of Defense Logistics Agency-Strategic Materials reports from 2011-2015. | GAO-16-161.

[a]Defense Logistics Agency–Strategic Materials specifically identified high purity yttrium.

Figure 4. Department of Defense Shortfall Risks Identified from 2011 to 2015.

DLA-Strategic Materials may identify a shortfall risk but not take mitigating actions for various reasons. For example, erbium and thulium had shortfall risks in 2013, but DLA-Strategic Materials did not recommend stockpiling them due to limited defense applications; limited data; and prioritization of more important materials, such as yttrium. In another instance, lanthanum had a shortfall risk in 2015, but DLA-Strategic Materials did not recommend it for stockpiling due to the availability of the material from companies in Australia, with which the United States has an agreement for priority delivery for defense needs.[16]

MIBP Identified Some Risks Related to Ensuring a Secure Supply of Rare Earths but Its Mitigation Strategy Is Reactive

MIBP also took actions to address supply disruption risks by identifying the risk of unavailability of heavy rare earths—those that are generally less abundant and more expensive due to their scarcity—used in defense applications. Heavy rare earths are at risk for a supply disruption due to the concentration of production outside the United States, which MIBP described in OUSD(AT&L)'s 2013 *Annual Industrial Capabilities Report to Congress*, the last one completed at the time of our review. MIBP based its analysis, in part, on public and industry supply and demand forecasts for rare earths and interviews with federal officials, subject matter experts, military service officials, and industry representatives, among other sources, according to officials.

To mitigate the risk, MIBP's strategy relies on the market: (1) to find new sources of rare earths driven by increasing demand, (2) to discover substitutes for rare earths, and (3) to support initiatives for recycling rare earths from waste products. MIBP's strategy is reactive, that is waiting for a supply disruption to occur and relying on the market to respond. Moreover, it may not be realistic. First, industry experts said that new mines and processing facilities are not economically viable as prices for rare earths have decreased and remain low. Second, Army officials, weapon systems subject matter experts, and industry representatives said that substitutes for some rare earths are not preferred or do not exist at this time. Third, while stating in 2012 that recycling initiatives could mitigate some supply risks, DOD determined that recycling is not economically feasible in 2014.[17]

In addition, MIBP does not have internal controls necessary to implement and monitor its strategy. Specifically, it has not established metrics to assess the extent to which its strategy could have mitigated the risk for a potential supply disruption of heavy rare earths. Nor has it defined the terms "reliable sources" and "secure supply" that are necessary to establish such metrics according to internal controls that call for identifying acceptable risk tolerances for what is reliable and secure. According to DOD officials, potential measures could be past performance history, including timeliness of delivery, or the financial health of companies in the rare earths supply chain. Further, MIBP has not defined the metrics for evaluating the extent of risk and the effectiveness of its strategy because, as indicated earlier it relies on the market to respond to a supply disruption. Without such metrics, it will be difficult for MIBP to monitor and adjust its mitigating actions.

In addition to its stated strategy, MIBP officials said to identify potential supply disruption risks for materials, such as rare earths, they rely on the military departments and program offices to elevate issues so that MIBP can react to them as they "bubble up." Similarly in our 2008 report on DOD's industrial policy, we found that DOD used an informal approach to identify supplier-base concerns, often relying on the military services, program offices, or prime contractors to identify and report these concerns, including gaps or potential gaps. We further found that as there was no requirement for when to report such gaps to higher-level offices, knowledge of defense supplier-base gaps across DOD could be limited.[18] Currently, MIBP partly relies on the Joint Industrial Base Working Group, which consists of department-wide representatives, to identify industrial base issues, such as rare earths unavailability or supply risk. We found that no guidance exists for when military departments and program offices are to report rare earth supply disruption risks, but MIBP officials reported that rare earths have never come up as an issue in the working group.

DOD Has Not Analyzed the Effects of a Supply Disruption on National Defense

The SMPB, chaired and vice-chaired by MIBP and DLA-Strategic Materials, has not taken a broader approach to comprehensively address the underlying risks of the unavailability of rare earths, including determining the effects and possible mitigating actions of the unavailability of rare earths, by leveraging the dispersed expertise that exists in the department. The current approach is thus fragmented. For example, although MIBP has offered a strategy to ensure a reliable supply, SMPB has not taken the additional step of assessing and building upon MIBP's work to develop and recommend a department-wide strategy. Further, although DLA-Strategic Materials' stockpiling efforts are proactive, according to DOD officials, the department has never used the rare earths from the National Defense Stockpile and does not know the time and cost to establish the capability to process the stockpiled material into a usable form, a necessary step to implement mitigating actions. Currently, there is a limited commercial capability in the United States for processing certain types of rare earths used in defense applications, according to government and industry officials. We previously reported in April 2010 that rebuilding a domestic rare earths supply chain may take significant time and money.[19] Industry representatives we spoke with during this review agreed that while there is domestic expertise, with limited domestic capability, turning certain stockpiled rare earths into a useable form would most likely have to be conducted outside the United States, which may also introduce risks of a supply disruption. SMPB has also not built upon DLA-Strategic Materials work to understand what is necessary to use the stockpiled materials.

According to defense officials, complete data on the use of rare earths in the defense industrial base are limited and challenging to collect, which can limit assessments of the defense supply chain needed to comprehensively address risks. For example, the Defense Contract Management Agency–Industrial Analysis Center, which conducts analyses of the industrial base for DOD, surveyed prime contractors and select subcontractors of acquisition category I programs—with a procurement cost of more than $2.79 billion—about their supply chains for rare earths.[20] According to Industrial Analysis Center officials, only 10 percent of surveys sent to prime contractors for 79 acquisition category I programs in July 2014 were returned with meaningful information. The prime contractors reported that among the reasons for not responding were concerns about the release of sensitive information and non-disclosure agreements with suppliers.

Although officials from MIBP and DLA-Strategic Materials also told us that their offices do not have the resources to track rare earths through defense supply chains from ore to weapon system, the SMPB has not leveraged the department's dispersed expertise in order to analyze the impact of unavailability on national defense. We found DOD officials who understood the vulnerability and effect of unavailability of rare earths based on their knowledge of the supply chain for select weapon systems. For example, a DOD-wide subject matter expert in radars at Naval Surface Warfare Center–Crane told us that rare earth magnets enable microwave tubes to effectively focus the radar's energy. Without the magnets, the radars cannot function and provide the needed capability to the warfighter. Further, he said that there are no substitutes for the rare earths in these magnets. His assessment was based, in part, on his knowledge of and relationships with sub-tier contractors and weapon system program officials. See figure 5 for a graphic example of how rare earths enter the defense supply chain for radars based on information from a defense industrial base report on radars, a subject matter expert and other DOD officials, and company representatives. He added that after the potential unavailability of rare earths in 2010, the Navy made a deliberate effort to understand the materials, vendors, and supply chains for DOD radars in order to determine the impact if rare earths were unavailable. We found the Navy's efforts are similar to some of the company approaches, discussed earlier, for managing risk. DLA-Strategic Materials officials later said that it is not always necessary to track a material through the supply chain to the weapon system to make a stockpiling recommendation but did so when beneficial. For example, for scandium used in ships, DLA-Strategic Materials did conduct a study that went from mine to platform and included program offices and prime contractors in their research due to specific manufacturing issues.

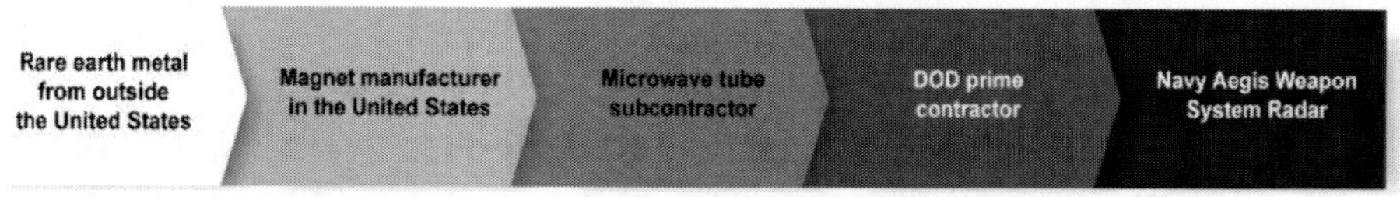

Source: DOD report and interviews with DOD officials and company representatives. | GAO-16-161.

Figure 5. Example of Rare Earth Materials in the Radar Supply Chain; Rare earth material magnets enable microwave tubes to effectively focus the radar's energy.

Federal internal control standards state that management should define objectives clearly to enable the identification of risks and define risk tolerances in specific and measurable terms, including defining what is to be achieved, who is to achieve it, how it will be achieved, and time frames for achievement.[21] Ultimately, SMPB has a goal of ensuring availability of rare earths that are critical to national security. DOD has taken actions without knowing the extent of the underlying risks the unavailability of rare earths have on its weapon systems specifically and national defense generally. By taking a comprehensive approach to meet this goal, such as addressing processing of stockpiled materials from identified shortfalls and defining metrics to measure risk and mitigating actions, the SMPB may be better positioned to guide other DOD offices when to report rare earths availability issues and to analyze the impact of the unavailability of the materials.

CONCLUSION

Reliable access to the material it needs, such as rare earths, is a bedrock requirement for DOD. Accordingly, a department-wide approach that facilitates DOD focusing its resources on those rare earths that are critical to national security would enable the department to meet its needs before supply for a specific rare earth becomes an emergency. However, DOD has no comprehensive, department-wide approach to determine which rare earths are critical to national security, and how to deal with potential supply disruptions to ensure continued, reliable access. It is important that such an approach includes, for example, leveraging the dispersed expertise in the department to identify which rare earths are critical, analyzing risks to their supply and the effects of a potential disruption, and taking actions to mitigate the specific risks. Developing a comprehensive approach for ensuring a sufficient supply of rare earths for national security needs—one that can establish criticality, assess supply risks, and identify mitigating actions—would better position DOD to help ensure continued functionality in weapon system components should a disruption occur, even though supply disruptions in rare earths have not occurred over the last several years. SMPB, with its statutory responsibility and leadership from MIBP and DLA-Strategic Materials, is well-positioned to take the lead on such an approach.

There is not yet agreement on what constitutes "critical" rare earths. While various organizations' definitions of critical may be similar, DOD has identified 15 of the 17 rare earths as critical over the last 5 years. This makes establishing priorities to analyze supply risk difficult. DLAStrategic Materials does a methodical job of analyzing risks for all materials, but its focus is a four-year timeframe with stockpiling as its mitigation tool. MIBP relies on other DOD organizations to identify and elevate risks, relies primarily on the market to resolve supply disruptions, and has not put in place measures to evaluate the success of its mitigating actions. The efforts by the military departments, such as the Navy, to identify and mitigate risks by monitoring the defense supplier base have been limited in scope or lacked department-wide involvement. Without clear definitions and metrics to identify associated risk tolerances, it will be difficult for MIBP to monitor and adjust its mitigating actions.

RECOMMENDATIONS FOR EXECUTIVE ACTION

To fully identify and mitigate risks associated with the availability of rare earths, we recommend that the Secretary of Defense take the following three actions:

- Direct the SMPB to designate which, if any, rare earths are critical to national security in order to provide a common DOD understanding of those materials and focus resources.
- Direct the SMPB to analyze the effect of unavailability of rare earths designated as critical to national security and develop a strategy to help ensure a secure supply for those designated critical to national security.
- Direct MIBP to define reliable sources and secure supply for rare earths in measurable terms and to provide metrics to determine the effectiveness of its actions to better ensure continued availability.

AGENCY COMMENTS AND OUR EVALUATION

We provided a draft of this report to DOD for review and comment. DOD concurred with all our recommendations. DOD and the United States Geological Survey also provided technical comments that we incorporated into the report as appropriate.

DOD concurred with our first two recommendations to the Secretary of Defense to direct the SMPB to designate which rare earths, if any, are critical to national security and to analyze the effects of the unavailability of rare earths critical to national security. In DOD's written response it identified that the SMPB will convene a meeting with the purpose of accomplishing both these tasks before the end of fiscal year 2016.

DOD also concurred with our recommendation to the Secretary of Defense to direct MIBP to define reliable sources and secure supply for rare earths and provide metrics to determine the effectiveness of its actions. In response DOD wrote that MIBP addresses this recommendation by carrying out its responsibilities under 10 U.S.C. § 139c by providing its input to the Secretary of Defense's *Biennial Report on Stockpile Requirements* and that the metrics and analyses used in *Biennial Report on Stockpile Requirements* do evaluate the effectiveness of the department's actions. However, we are concerned whether the *Biennial Report on Stockpile Requirements* addresses the department's responsibilities under 10 U.S.C. § 139c. As stated earlier in the report, in carrying out responsibilities under 50 U.S.C. § 98 et seq., DLA-Strategic Materials officials caveat their analyses as being valid for a single point in time and for decisions related to the National Defense Stockpile and not for other statutory responsibilities such as MIBP's identification of materials critical to national security in 10 U.S.C. § 139c. Further, MIBP, in its response, has not explained how its input for the *Biennial Report on Stockpile Requirements* will define reliable sources and secure supply, and what metrics that the report will provide to determine the effectiveness of the department's mitigation efforts.

In addition, DOD, in its technical comments, disagreed with our characterization that DOD has no department-wide approach for critical materials. We agree that DLA-Strategic Materials' process may be ongoing, methodical, and incorporate information from DOD, government, and industry sources, however, as stated in the report, the assessments are limited to a four-year time frame related to DOD defined national emergencies. Further, during our review, DLA-Strategic Materials officials told us that they caveat their analyses as being valid for a single point in time and for decisions specifically related to the National Defense Stockpile, and thereby not necessarily for determining criticality for national security purposes. We continue to believe our findings demonstrate the need for a comprehensive approach for ensuring a sufficient supply of rare earths for national security needs—one that can establish criticality, assess supply risks, and identify mitigating actions—beyond planning scenarios used for purposes of the National Defense Stockpile. Such an approach would better position DOD to help ensure continued functionality in weapon system components should a disruption occur.

Marie A. Mak
Director
Acquisition and Sourcing Management

APPENDIX I: OBJECTIVES, SCOPE, AND METHODOLOGY

The Joint Explanatory Statement accompanying the Carl Levin and Howard P. "Buck" McKeon National Defense Authorization Act (NDAA) for Fiscal Year 2015 included a provision for GAO to review the Department of

Defense's (DOD) efforts to identify and mitigate risks in its rare earth materials (rare earths) supply chain, including the Defense Logistics Agency-Strategic Materials' (DLA-Strategic Materials) stockpiling efforts. We assessed the extent that DOD (1) determined which rare earths, if any, are critical to national security; and (2) has identified and mitigated risks associated with rare earths, including the effects of a potential supply disruption.

To determine the extent DOD identified which rare earths, if any, were critical to national security, we reviewed DOD reports from 2011 to 2015 to understand the analyses and criteria that DOD conducted and used to determine critical rare earths. We selected 2011 to 2015, because we last reported on rare earths in 2010. We reviewed DOD reports from the Office of the Undersecretary of Defense for Acquisition, Technology, and Logistics including the biennial *Strategic and Critical Materials Report on Stockpile Requirements for the National Defense Stockpile* and the *Annual Industrial Capabilities Report* from 2011 through 2015 to determine the analyses DOD had conducted to identify which rare earths may be critical. We reviewed relevant legislation including the Strategic and Critical Materials Stock Piling Act (50 U.S.C. § 98 et seq.), 10 U.S.C. § 139c that outlines the responsibilities of the Office of the Deputy Assistant Secretary of Defense for Manufacturing and Industrial Base Policy (MIBP), and 10 U.S.C. § 187 that outlines the duties of the Strategic Materials Protection Board (SMPB). We also reviewed available meeting minutes from the SMPB from 2007 through 2015. Further we reviewed relevant DOD components' actions to determine if they had separately identified any critical rare earths in their supply chains. We interviewed DOD officials from MIBP, SMPB, and DLA-Strategic Materials, which is the program manager of the National Defense Stockpile. In addition, we collected data, documents, and conducted interviews with officials from the DOD components, such as Defense Procurement and Acquisition Policy and the Office of the Assistant Secretary of Defense for Logistics and Materiel Readiness, the military departments—Air Force, Army, and Navy—and their respective research laboratories, and members of the Joint Industrial Base Working Group. We collected data supporting DLA-Strategic Materials assessment of materials from 2011 to 2015, documents, and conducted interviews with officials at other offices and agencies that had knowledge of rare earths, the defense industrial base, and the National Defense Stockpile including: Office of the Inspector General; Defense Logistics Agency, Warstopper Program; Congressional Research Service; the Department of the

Interior, United States Geological Survey; Department of Commerce, Bureau of Industry and Security; and the Department of Energy.

To assess the extent to which DOD has identified risks and taken action to mitigate the risks associated with the availability of rare earths, and analyzed the effects of a potential supply disruption, we reviewed DOD reports from 2011 to 2015 to understand the analyses DOD had conducted to identify risks related to the availability of rare earths and any mitigating actions. We reviewed DOD reports from the Office of the Undersecretary of Defense for Acquisition, Technology, and Logistics including the biennial *Strategic and Critical Materials Report on Stockpile Requirements for the National Defense Stockpile* and the *Annual Industrial Capabilities Report* from 2011 through 2015 to determine the analyses DOD had conducted to identify and mitigate risks from rare earths. We reviewed relevant legislation including the Strategic and Critical Materials Stock Piling Act (50 U.S.C. § 98 et seq.), 10 U.S.C. § 139c that outlines the responsibilities of MIBP, and 10 U.S.C. § 187 that outlines the duties of the SMPB. In addition, we reviewed applicable policies, such as DOD Instruction 5000.60, *Defense Industrial Base Assessments* (July 18, 2014), to determine additional requirements for DOD in managing risks in its rare earths supply chain. We also reviewed available meeting minutes from the SMPB from 2007 through 2015. Further we reviewed relevant DOD components' actions to determine if they had separately identified any risks from rare earths in their supply chains and taken mitigating actions. We interviewed DOD officials from MIBP, SMPB, and DLA-Strategic Materials, the program manager of the National Defense Stockpile. We also collected documents and conducted interviews with the Institute for Defense Analyses, a not-for-profit corporation that operates three federally funded research and development centers, that conducts research and created and operates the econometric model for DLA-Strategic Materials; and with Oak Ridge National Laboratory, a federally funded research and development center that developed the Strategic Materials Analysis and Reporting Topography software tool. In addition, we collected data, documents, and conducted interviews with officials from the DOD components, such as Defense Procurement and Acquisition Policy and the Office of the Assistant Secretary of Defense for Logistics and Materiel Readiness, the military departments—Air Force, Army, and Navy—and their respective research laboratories, and members of the Joint Industrial Base Working Group. We collected data, documents, and conducted interviews with officials at other offices and agencies that had knowledge of rare earths, the defense industrial base, and the National Defense Stockpile including:

Defense Contract Management Agency–Industrial Analysis Center; Office of the Inspector General; Defense Logistics Agency, Warstopper Program; Congressional Research Service; the Department of the Interior, United States Geological Survey; Department of Commerce, Bureau of Industry and Security; and the Department of Energy.

We conducted interviews and collected information for a non-generalizable sample of 20 cases for illustrative purposes on rare earths and materials risk management that included:

- Three DOD offices (Naval Surface Warfare Center–Crane both Radars and Lasers, and Army Program Executive Office for Ammunition) with purview over at least seven weapon systems selected on rare earths in the weapon systems;
- Three vendors of rare earth components selected on rare earths in their products;
- Two rare earths companies selected on rare earths in their products;
- One trade association selected on its involvement in the rare earth industry;
- Six individuals selected on their knowledge of rare earths;
- Three defense contractors selected on their defense products and materials risk management processes; and
- Two companies, including a petroleum company, selected on their materials risk management processes.

We reviewed DOD's sources of data for its analytical processes. Based on our review, we determined that the information that we collected were sufficiently reliable for our report. We assessed DOD's policies, procedures, and practices against criteria in applicable statutes, *Standards for Internal Control in the Federal Government*, and GAO's framework for risk assessment.[1]

We conducted this performance audit from February 2015 to February 2016 in accordance with generally accepted government auditing standards. Those standards require that we plan and perform the audit to obtain sufficient, appropriate evidence to provide a reasonable basis for our findings and conclusions based on our audit objectives. We believe that the evidence obtained provides a reasonable basis for our findings and conclusions based on our audit objectives.

End Notes

[1] Rare earth materials refer to items that contain one or more of 17 similar metals, which have unique properties, such as magnetism at high temperatures. The rare earths are an International Union of Pure and Applied Chemistry named group.

[2] GAO, *Rare Earth Materials in the Defense Supply Chain,* GAO-10-617R (Washington, D.C.: Apr. 14, 2010).

[3] Ike Skelton National Defense Authorization Act for Fiscal Year 2011, Pub. L. No. 111-383 § 843; 10 U.S.C. § 187; 10 U.S.C. § 139c and 50 U.S.C. § 98 et seq. GAO, *Standards for Internal Control in the Federal Government*, GAO/AIMD-00-21.3.1 (Washington, D.C.: Nov. 1999). *Risk Management: Further Refinements Needed to Assess Risks and Prioritize Protective Measures at Ports and Other Critical Infrastructure,* GAO-06-91 (Washington, D.C.: Dec. 15, 2005).

[4] According to the United States Geological Survey, the heavy rare earth elements include yttrium, terbium, dysprosium, holmium, erbium, thulium, ytterbium, and lutetium. United States Geological Survey, *Preliminary Estimates of the Quantities of Rare-Earth Elements Contained in Selected Products and in Imports of Semimanufactured Products to the United States, 2010*, Open-File Report 2013–1072 (Reston, Va.: 2013).

[5] Congressional Research Service, *Rare Earth Elements in National Defense: Background, Oversight Issues, and Options for Congress*, R41744 (Dec. 23, 2013).

[6] DOD, Office of the Under Secretary of Defense (Acquisition, Technology and Logistics), *Report of Meeting Department of Defense Strategic Materials Protection Board* (Dec. 12, 2008).

[7] "Critical" and "strategic and critical" are terminology identified in statute. See table 3 for the definitions.

[8] U.S. Department of Defense, Inspector General, *Procedures to Ensure Sufficient Rare Earth Elements for the Defense Industrial Base Need Improvement*, DODIG-2014-091 (Alexandria, Va.: July 3, 2014).

[9] GAO, *Department of Defense: A Departmentwide Framework to Identify and Report Gaps in the Defense Supplier Base Is Needed,* GAO-09-5 (Washington, D.C.: Oct. 7, 2008).

[10] Ike Skelton NDAA for Fiscal Year 2011, Pub. L. No. 111-383, § 843 (identification of critical rare earth materials in response to statutorily-required DOD assessment of supply and demand for critical rare earth materials in defense applications, identifying which, if any, rare earth material meets both of the criteria listed in table 3 here for the Under Secretary of Defense for AT&L); 10 U.S.C. § 187(e); 10 U.S.C. § 139c(d); and 50 U.S.C. § 98h-3(1).

[11] GAO considers programs or activities to be fragmented when more than one federal agency (or more than one organization within an agency) is involved in the same broad area of national need, which may result in inefficiencies in how the government delivers services. GAO, *Fragmentation, Overlap, and Duplication: An Evaluation and Management Guide,* GAO-15-49SP (Washington, D.C.: Apr. 14, 2015).

[12] GAO developed a framework for risk management based on industry best practices and other criteria, such as past GAO work and feedback from academic experts in risk management. The risk management framework is divided into five major phases: (1) setting strategic goals and objectives, and determining constraints; (2) assessing the risks; (3) evaluating alternatives for addressing these risks; (4) selecting the appropriate alternatives; and (5) implementing the alternatives and monitoring the progress made and results achieved. GAO-06-91. GAO, *Risk Management: Strengthening the Use of Risk Management Principles in Homeland Security,* GAO-08-904T (Washington, D.C.: June 25, 2008).

[13] The SMPB, in December 2008, reported only beryllium met its definition of a material critical to national security, that is, a strategic material for which (1) DOD dominates the market for the material; (2) DOD's full and active involvement and support are necessary to sustain and shape the market; and (3) there is significant and unacceptable risk of supply disruption. The House Report accompanying the NDAA for Fiscal Year 2010 expressed concern about the narrow limits of SMPB's definition. H.R. Rep. No. 111-168, at 350 (2010). The Ike Skelton NDAA for Fiscal Year 2011 added a new definition of materials critical to national security for SMPB purposes to mean those materials upon which the production or sustainment of military equipment is dependent and the supply of which could be restricted by actions or events outside the control of the government of the United States. Pub. L. No. 111-383, § 829. To date, the SMPB has not designated any materials as critical to national security.

[14] GAO/AIMD-00-21.3.1.

[15] DOD, Under Secretary of Defense for Acquisition, Technology and Logistics, *Strategic and Critical Materials 2015 Report on Stockpile Requirements*, 9-05A8E24 (Jan. 2015).

[16] The bilateral Security of Supply arrangement between Australia and the United States allows DOD to request priority delivery for DOD contracts, subcontracts, or orders from companies in Australia. Similarly, the arrangement allows Australia to request priority delivery for their contracts and orders with U.S. companies.

[17] Department of Defense, Office of the Under Secretary of Defense for Acquisition, Technology, and Logistics, *Diversification of the Supply Chain and Reclamation Activities Related to Rare Earths*, 7-A4D556E (Feb. 2014).

[18] GAO-09-5.

[19] GAO-10-617R.

[20] According to DOD Instruction 5000.02-*Operation of the Defense Acquisition System* (Jan. 7, 2015), acquisition category I programs are major defense acquisition programs with a dollar value for all increments of the program estimated by the Defense Acquisition Executive to require an eventual total expenditure for research, development, and test and evaluation of more than $480 million or, for procurement, of more than $2.79 billion in fiscal year 2014 constant dollars.

[21] GAO/AIMD-00-21.3.1.

End Note for Appendix I

[1] Ike Skelton National Defense Authorization Act for Fiscal Year 2011, Pub. L. No. 111-383 § 843; 10 U.S.C. § 187; 10 U.S.C. § 139c and 50 U.S.C. § 98 et seq. GAO, *Standards for Internal Control in the Federal Government*, GAO/AIMD-00-21.3.1 (Washington, D.C.: Nov. 1999). *Risk Management: Further Refinements Needed to Assess Risks and Prioritize Protective Measures at Ports and Other Critical Infrastructure,* GAO-06-91 (Washington, D.C.: Dec. 15, 2005).

In: Rare Earth Elements
Editor: Kristin Waters

ISBN: 978-1-53610-120-1

Chapter 2

RARE EARTH ELEMENTS IN NATIONAL DEFENSE: BACKGROUND, OVERSIGHT ISSUES, AND OPTIONS FOR CONGRESS*

Valerie Bailey Grasso

SUMMARY

Some Members of Congress have expressed concern over U.S. acquisition of rare earth materials composed of rare earth elements (REE) used in various components of defense weapon systems. Rare earth elements consist of 17 elements on the periodic table, including 15 elements beginning with atomic number 57 (lanthanum) and extending through number 71 (lutetium), as well as two other elements having similar properties (yttrium and scandium). These are referred to as "rare" because although relatively abundant in total quantity, they appear in low concentrations in the earth's crust and extraction and processing is both difficult and costly.

From the 1960s to the 1980s, the United States was the leader in global rare earth production. Since then, production has shifted almost entirely to China, in part due to lower labor costs and lower environmental standards. Some estimates are that China now produces about 90- 95% of the world's rare earth oxides and is the majority producer of the world's two strongest magnets, samarium cobalt (SmCo)

* This is an edited, reformatted and augmented version of a Congressional Research Service, Publication No. R41744, dated December 23, 2013.

and neodymium iron boron (NeFeB) permanent, rare earth magnets. In the United States, Molycorp, a Mountain Pass, CA mining company, recently announced the purchase of Neo Material Technologies. Neo Material Technologies makes specialty materials from rare earths at factories based in China and Thailand. Molycorp also announced the start of its new heavy rare earth production facilities, Project Phoenix, which will process rare earth oxides from ore mined from the Mountain Pass facilities.

In 2010, a series of events and press reports highlighted what some referred to as the rare earth "crisis." Some policymakers were concerned that China had cut its rare earth exports and appeared to be restricting the world's access to rare earths, with a nearly total U.S. dependence on China for rare earth elements, including oxides, phosphors, metals, alloys, and magnets. Additionally, some policymakers had expressed growing concern that the United States had lost its domestic capacity to produce strategic and critical materials, and its implications for U.S. national security.

Pursuant to Section 843, the Ike Skelton National Defense Authorization Act for FY2011 (P.L. 111-383) and S.Rept. 111-201 (accompanying S. 3454), Congress had mandated that the Secretary of Defense conduct an assessment of rare earth supply chain issues and develop a plan to address any vulnerabilities. DOD was required to assess which rare earths met the following criteria: (1) the rare earth material was critical to the production, sustainment, or operation of significant U.S. military equipment; and (2) the rare earth material was subject to interruption of supply, based on actions or events outside the control of the U.S. government. The seven-page report was issued in March 2012.

In October 2013, DOD released its Annual Industrial Capabilities Report to Congress in accordance with Section 2504 of Title 10, United States Code (U.S.C.). The report states that, due to global market forces, the overall demand for rare earth materials has decreased, as prices for most rare earth oxides and metals have declined since 2011.

Given DOD's assessment of the supply and demand for rare earths for defense purposes, coupled with the recent announcement of Molycorp's proposed acquisition of Neo Material Technologies, Congress may choose to use its oversight role to seek more complete answers to the following important questions:

- Given Molycorp's purchase of Neo Material Technologies and the potential for the possible migration of domestic rare earth minerals to Molycorp's processing facilities in China, how may this move affect the domestic supply of rare earth minerals for the production of U.S. defense weapon systems?
- Given that DOD's assessment of future supply and demand was based on previous estimates using 2010 data, could there be new

concern for a possible rare earth material supply shortage or vulnerability that could affect national security?

- Are there substitutes for rare earth materials that are economic, efficient, and available?
- Does dependence on foreign sources alone for rare earths pose a national security threat?

Congress may encourage DOD to develop a collaborative, long-term strategy designed to identify any material weaknesses and vulnerabilities associated with rare earths and to protect long-term U.S. national security interests.

INTRODUCTION

This report discusses rare earth elements used in Department of Defense (DOD) weapon systems, current problematic oversight issues, and options for Congress to consider. Rare earth elements (also referred to as REEs and by the shorthand term "rare earths") include the lanthanide series of 15 elements on the periodic table, beginning with atomic number 57 (lanthanum) and extending through element number 71 (lutetium). Two other elements, yttrium and scandium, often occur in the same rare earth deposits and possess similar properties.[1] These 17 elements are referred to as "rare" because while they are relatively abundant in quantity, they appear in low concentrations in the earth's crust and economic extraction and processing is both difficult and costly.

The United States is a major consumer of products containing rare earth elements. These elements are incorporated into many sophisticated technologies with both commercial and defense applications. From the 1960s to the 1980s, the United States was the leader in global production of rare earths. Since that time, processing and manufacturing of the world's supply of rare earths and downstream value-added forms such as metals, alloys, and magnets have shifted almost entirely to China, in part due to lower labor costs and lower environmental standards.

A series of events and ensuing press reports have highlighted the rare earth "crisis," as some refer to it. One such event occurred in July 2010, when China's Ministry of Commerce announced that China would cut its exports of rare earth minerals by about 72%. In September 2010, China temporarily cut rare earth exports to Japan, apparently over a maritime dispute. This dispute highlighted the potential for disruption of the world's supply of rare earth materials.

Some Members of Congress are concerned with the potential for a nearly total U.S. dependence on foreign sources for rare earth elements and the implications of this dependence for national security. Congress has been interested in the rare earth issue largely because:

- the world is almost wholly dependent on a single national supplier—China—for rare earths;
- the United States has no production of some heavy rare earths (terbium to lutetium and yttrium);[2]
- the United States has little production of rare earth metals, powders, and NeFeB magnets;
- there may be repercussions if these materials are not available for commercial and defense applications; and
- the rare earths supply chain vulnerability question may adversely affect the ability of the United States to plan strategically for its national security needs.

In April 2010 Congress required the Government Accountability Office (GAO) to examine rare earths in the defense supply chain and also required the Secretary of Defense to assess the defense supply chain and develop a plan to address any shortfalls or other supply chain vulnerabilities, including a specific requirement to present a plan for the restoration of domestic NeFeB magnet production. GAO concluded that revamping the defense supply chain could take 15 years or more. Congress has required that the Secretary of Defense, pursuant to the Ike Skelton National Defense Authorization Act for FY2011 (P.L. 111-383), conduct an assessment of the rare earths supply chain issues and develop a plan to address any supply chain vulnerabilities. DOD's report was released in March 2012.

Congress may want answers to at least four important questions on rare earth elements: (1) Are rare earth elements essential to U.S. national security? (2) How would a scarcity of rare earths affect the delivery or performance of defense weapon systems? (3) Is the United States vulnerable to supply disruptions, and if so, are there readily available and equally effective substitutes? (4) What are the short-term and long-term options that DOD may consider in response to a lack of domestic rare earth element production and China's continued dominance?

MAJOR NEW DEVELOPMENTS

H.R. 1960, National Defense Authorization Act (NDAA) for FY2014

H.R. 1960 was introduced in the House on May 14, passed the House in a recorded vote (315- 108) on June 14, and was referred to the Senate on July 8, 2013. The bill contains several provisions which would, if enacted into law, give the President more authority to conserve strategic and critical materials, as well as direct the Secretary of Defense to report on plans to assess the supply chain diversification for rare earth substitutes and develop risk mitigation strategies. The provisions appear below.

Subtitle B—National Defense Stockpile

Section 1411. Use of National Defense Stockpile for the Conservation of a Strategic and Critical Materials Supply

Section 1411 would modify the President's authority to maintain and manage a national defense stockpile, and allow the Defense Logistics Agency to more proactively engage in the market. These changes would grant the President the authority to conserve strategic and critical materials.

(a) Presidential Responsibility for Conservation of Stockpile Materials - Section 98e(a) of Title 50, United States Code, is amended:

(1) by redesignating paragraphs (5) and (6) as paragraphs (6) and (7), respectively; and

(2) by inserting after paragraph (4) the following new paragraph (5):

"(5) provide for the recovery of any strategic and critical material from excess materials made available for recovery purposes by other Federal agencies;"

(b) Uses of National Defense Stockpile Transaction Fund - Section 98h(b)(2) of Title 50, United States Code, is amended—

(1) by redesignating subparagraphs (D) through (L) as subparagraphs (E) through (M), respectively; and

(2) by inserting after subparagraph (C) the following new subparagraph (D): "(D) Encouraging the conservation of strategic and critical materials."

(c) Development of Domestic Sources - Section 98h-6(a) of Title 50, United States Code, is amended, in the matter preceding paragraph (1), by inserting 'and conservation' after 'development'.

Section 1412. Authority to Acquire Additional Materials for the National Defense Stockpile

Section 1412 would provide authority to acquire certain additional strategic and critical materials for the National Defense Stockpile. The materials anticipated to be acquired have been identified to meet the military, industrial, and essential civilian needs of the United States.

(a) Acquisition Authority - Using funds available in the National Defense Stockpile Transaction Fund, the National Defense Stockpile Manager may acquire the following materials determined to be strategic and critical materials required to meet the defense, industrial, and essential civilian needs of the United States:

(1) Ferroniobium.

(2) Dysprosium Metal.

(3) Yttrium Oxide.

(4) Cadmium Zinc Tellurium Substrate Materials.

(5) Lithium Ion Precursors.

(6) Triamino-Trinitrobenzene and Insensitive High Explosive Molding Powders.

(b) Amount of Authority - the National Defense Stockpile Manager may use up to $41,000,000 of the National Stockpile Transaction Fund for acquisition of the materials specified in subsection (a).

(c) Fiscal Year Limitation - The authority under this section is available for purchases during Fiscal Year 2014 through Fiscal Year 2019.

Directive Report Language

In H.R. 1960, under Title XVI, Industrial Base Matters, there are two reporting requirements required by the House Armed Services Committee that address congressional concerns over maintaining a secure access and a diverse supply chain for rare earth elements to be used for national security purposes and in defense weapon systems.

The first directive requires the Under Secretary of Defense for Acquisition, Technology and Logistics to submit a report to the congressional defense committees, by February 1, 2014, to outline a risk mitigation strategy focused on securing the necessary supplies of rare earth elements. The report language reads as follows:

Title XVI – Industrial Base Matters

Report on the Diversification of Supply Activities Related to Rare Earth Elements

The committee is aware that in response to the report required by section 843 of the Ike Skelton National Defense Authorization Act for Fiscal Year 2011 (P.L. 111-383) and based on forecasting demand for fiscal year 2013 only, the Under Secretary of Defense for Acquisition, Technology, and Logistics concluded that domestic production of rare earth elements could satisfy the level of consumption required to meet defense procurement needs by fiscal year 2013, with the exception of yttrium. However, the committee observes that the Future Years Defense Program indicates that consumption of rare earth elements is expected to increase after 2013. Specifically, the report on the feasibility and desirability of recycling, recovery, and reprocessing of rare earth elements required by the conference report (H.Rept. 112-329) to accompany the National Defense Authorization Act for Fiscal Year 2012, states that each SSN-774 *Virginia*-class submarine would require approximately 9,200 pounds of rare earth materials, each DDG-51 Aegis destroyer would require approximately 5,200 pounds of these materials, and each F-35 Lightning II aircraft would require approximately 920 pounds of these materials.

The committee is aware that the Department of Defense intends to pursue a three-pronged strategy to secure supplies of rare earth elements, which consists of diversification of supply, pursuit of substitutes, and a focus on reclamation of waste, as part of a larger U.S. Government recycling effort. The committee believes that diversification of supply activities related to rare earth elements is necessary in order to meet the growing demand for these materials, but the committee is concerned that some of these processes may prove to be technically difficult or so expensive that they are deemed cost-prohibitive.

Therefore, the committee directs the Under Secretary of Defense for Acquisition, Technology, and Logistics to submit a report to the congressional defense committees by February 1, 2014, on the Department's risk mitigation strategy for rare earth elements, which should include, at a minimum, the following elements:

(1) A list and description of the programs initiated or planned to reclaim rare earth elements by the Department, along with a description of the materials reclaimed or expected to be reclaimed from such programs;

(2) An assessment of the cost of materials produced by these reclamation efforts compared to the cost of newly-mined materials;

(3) An assessment of availability of reliable suppliers in the National Defense Industrial Base for the reclamation and reprocessing of rare earth elements;

(4) A list of alternative sources of supply, such as mine tailings, recycled components, and consumer waste, that the Department has investigated or plans to investigate;

(5) A physical description of alternative sources of supply with corresponding geologic characteristics, such as grade, resource size, and the amenability of that feedstock to metallurgical processing;

(6) A description of the materials that the Department plans to obtain via the Defense Priorities and Allocations System; and

(7) Other diversification of supply activities deemed relevant by the Under Secretary.[3]

The second directive requires DOD to perform an assessment of the potential for incorporating the substitution of non-rare earth materials into components of the Joint Strike Fighter, based on the supply chain challenges faced in securing components containing rare earth materials.

Report on the Implementation of Rare Earth Elements Strategy in the Joint Strike Fighter Program

> The committee is aware that the Department of Defense intends to pursue a three-pronged strategy to secure supplies of rare earth elements, which consists of diversification of supply, pursuit of substitutes, and a focus on reclamation of waste as part of a larger U.S. Government recycling effort. However, it remains unclear how this strategy will be implemented in the Department's major defense acquisition programs (MDAPs). Several high-profile MDAPs, including the F-35 Lightening II program, may use significant amounts of rare earth elements in full-rate production. The committee is concerned that the introduction of substitute materials and components may increase acquisition and sustainment costs through the qualification of manufacturers for substitutes, implementation of engineering changes to accommodate substitutes, and the long-term costs associated with supplier networks.
>
> Therefore, the committee directs the Assistant Secretary of the Navy for Research, Development and Acquisition, in coordination with the Program Executive Officer for the F-35, to submit a report to the congressional defense committees by February 15, 2014, on the potential for substitution of components and materials into F-35 aircraft to reduce

consumption of rare earth materials. The report, which may include a classified annex, should include the following:

(1) A list and description of subsystems that contain rare earth elements and the approximate quantities of each rare earth element by subsystem;

(2) An assessment of the potential to incorporate substitute components or materials in each subsystem based on technical acceptability, to include consideration of performance requirements, and engineering changes that may be necessary for integration of the substitute; and

(3) An assessment of the potential to incorporate substitute components or materials in each subsystem based on cost acceptability to include consideration of material costs, qualification and testing costs, and engineering change costs.[4]

S. 1197, NDAA for FY2014

S. 1197 was introduced on June 20, 2013, referred to the Armed Services Committee. The bill was considered for Senate floor action on December 9, 2013.

S. 1600, Critical Minerals Policy Act of 2013

S. 1600, the Critical Minerals Policy Act of 2013, was introduced on October 29, 2013, and referred to the Energy and Natural Resources Committee. The bill would require the Secretary of Interior and the Secretary of Energy to amend current policies, including "facilitate the reestablishment of domestic, critical mineral designation, assessment, production, manufacturing, recycling, analysis, forecasting, workforce, education, research, and international capabilities in the United States."[5]

H.R. 761, National Strategic and Critical Minerals Production Act of 2013

H.R. 761, the National Strategic and Critical Minerals Production Act of 2013 was introduced on February 15, 2013, and referred to the Committee on

Natural Resources on July 8, 2013, (H.Rept. 113-138). The bill passed in a recorded vote, 246-178, and was referred to the Senate Energy and Natural Resources Committee on December 19, 2013. The bill would require both the Secretary of the Interior and the Secretary of Agriculture to more efficiently develop domestic sources of the minerals and materials of strategic and critical importance to U.S. economic and national security, and manufacturing competitiveness.[6]

H.R. 981, RARE Act of 2013

H.R. 981, the RARE Act of 2013, was introduced on March 6, 2013, referred to the Subcommittee on Energy and Mineral Resources on March 7, 2013, and ordered to be reported by Unanimous Consent on May 15, 2013. The bill would require the Secretary of Interior to conduct an assessment of current global rare earth element resources and the potential future global supply.

H.R. 1063, National Strategic and Critical Minerals Policy Act of 2013

H.R. 1063, the National Strategic and Critical Minerals Policy Act of 2013, was introduced on March 12, 2013, referred to the Subcommittee on Energy and Mineral Resources on March 15, 2013, and ordered to be reported by Unanimous Consent on May 15, 2013. The bill would require the Secretary of the Interior to conduct an assessment of the current and future demands for the minerals critical to United States manufacturing, agricultural competitiveness, economic and national security.

President Obama's Announcement on the Joint World Trade Organization Dispute Resolution Case on China[7]

On March 13, 2012, President Obama made an announcement that the United States "had asked the World Trade Organization to facilitate formal consultations with China over its limits on rare-earth exports, in a case filed jointly with Japan and the European Union (EU)."[8] In announcing the case against China, the United States believes that China is illegally limiting exports of rare earths and pressuring foreign companies to move to China to take advantage of lower prices for domestic rare earth customers. In its defense, China claims that reducing the size of exports will help alleviate environmental hazards resulting from rare earth mining.[9]

The U.S. Trade Representative issued the following statement:

> America's workers and manufacturers are being hurt in both established and budding industrial sectors by these policies. China continues to make its export restraints more restrictive, resulting in massive distortions and harmful disruptions in supply chains for these materials throughout the global marketplace," said Ambassador Kirk. "The launch of this case against China today, along with the President's creation of the Interagency Trade Enforcement Center, reflects the Obama Administration's commitment to make all of our trading partners play by the rules. We will continue fighting for a level playing field for American workers and manufacturers in order to grow our economy, and ensure open markets for products made in America."
>
> The United States recently won a WTO challenge against China's export restraints on nine other industrial inputs. China's export restraint measures on rare earths, tungsten, and molybdenum appear to be part of the same troubling industrial policy aimed at providing substantial competitive advantages for Chinese manufacturers.
>
> China imposes several different types of unfair export restraints on the materials at issue in today's consultations request, including export duties, export quotas, export pricing requirements as well as related export procedures and requirements. Because China is a top global producer for these key inputs, its harmful policies artificially increase prices for the inputs outside of China while lowering prices in China. This price dynamic creates significant advantages for China's producers when competing against U.S. producers – both in China's market and in other markets around the world. The improper export restraints also contribute to creating substantial pressure on U.S. and other non-Chinese downstream producers to move their operations, jobs, and technologies to China.[10]

In July 2012, a WTO panel was convened. Some estimates were that the panel of judges will have about six months to complete their inquiry and then issue a final report. The WTO joint dispute resolution panel estimated that a final report would be issued in November 2013.[11] However, as of December 2013 no final report has been issued.

2013 Annual Industrial Capabilities Report to Congress

In October 2013, DOD released its Annual Industrial Capabilities Report to Congress. The reported stated that, for a variety of reasons, the overall supply chain issues regarding the availability of rare earth materials first raised in 2010 and 2011 have improved, as discussed here.

> Rare earth elements constitute a group of materials with numerous commercial as well as defense applications. These materials gained considerable attention in 2011 as prices increased drastically and concerns rose over their availability especially due to one nation, China, being the source of over 95 percent of the global supply. However, global market forces are leading to positive changes in rare earth supply chains and a sufficient supply of most of these materials likely will be available to the defense industrial base. In fact, there has been a significant reversal in the situation in 2012 from the prevailing market conditions in 2011. Overall demand for rare earth materials has decreased considerably from the previous year. A leading global expert forecast in August 2012 that 2012 global demand is expected to be 20 percent less than forecast in early 2011. Similarly, this expert believes that demand in 2016 will be 20 percent less than forecast in 2011. One factor contributing to reduced demand is the substitution of other materials for rare earth materials. There has also been reduced rare earth usage in individual applications and a drawdown in inventories accumulated the previous year. In conjunction with these factors, there has been an increase in supply of material from outside of China. As a result, prices for most rare earth oxides and metals have declined approximately 60 percent from their peaks in the summer of 2011.
>
> A strengthening and diversification of the supply chain for rare earth materials, including in the U.S., is anticipated in the coming year. The private sector's reaction to market forces has been to increase exploration for rare earth materials and development of downstream processing capabilities. According to one industry expert, there are over 400 rare earth projects under review globally, approximately four dozen of which may be considered in advanced stages of development in over a dozen countries worldwide. Two new projects in particular should have a significant impact on markets in 2013. Rare earth oxide production that has commenced in the U.S. and Malaysia based on mined material from the U.S. and Australia, respectively, will add approximately 40,000 tons of oxide production capacity to global supply in 2013, close to one-third of forecasted demand. Furthermore, the facility in the U.S. will have the capability to increase its capacity by an additional 20,000 tons in 2013.[12]

2013 Strategic and Critical Materials Report on Stockpile Requirements

The Secretary of Defense is required to report to Congress, on a biannual basis, recommendations on defense planning and projections for stockpile requirements (the availability of supplies of strategic and critical materials) under certain national emergency planning assumptions.[13] This report outlines the current state of strategic and critical materials for defense use, and the proposed mitigation strategies. These strategies include stockpiling, substitution with other materials, export reductions, and increasing procurement.[14] This report found that of the 76 materials evaluated for the potential for shortfalls (as defined as insufficiently reliable production to meet demands), 23 materials were found to represent a potential shortfall. This study included all rare earth elements, except promethium. Six rare earth element shortfalls were identified: they are: yttrium, dysprosium, erbium, terbium, thulium, and scandium.[15]

Department of Defense Report Responding to Section 843

DOD released a seven-page report in March 2012.[16] The report stated that "Seven of the 17 rare earth elements were found to meet the criteria established in Section 843."[17] They are dysprosium, erbium, europium, gadolinium, neodymium, praseodymium, and yttrium. DOD's assessment of the forecast for a domestic supply for key rare earths concluded: "by 2012 U.S. production (for seven rare earths used in military applications) could satisfy the level of consumption required to meet defense procurement needs, with the exception of yttrium."[18]

In an April 2012 interview with Bloomberg News, the DOD head of industrial policy confirmed that DOD uses less than 5% of rare earths used in the United States, and that DOD was closely monitoring the rare earth materials market for any projected shortfalls or failures to meet mission requirements. Brett Lambert, Deputy Assistant Secretary of Defense for Manufacturing and Industrial Base Policy, suggested that if material shortages were projected, DOD would seek congressional approval to stockpile materials. Other measures could include the use of contingency contracting to meet DOD requirements.[19]

Background on Rare Earth Elements

What Are Rare Earth Elements?

According to the U.S. Geological Survey (USGS),[20] there are 17 rare earth elements on the periodic table. The first 15 elements begin with atomic number 57 (lanthanum) and extend through element number 71 (lutetium); two other elements, yttrium and scandium, have similar properties. Rare earths are not particularly rare but are found in low concentrations in the earth's crust. The economics of locating and retrieving them are challenging. Rare earths are divided into two groups: light rare earth elements (LREE) - lanthanum, cerium, praseodymium, neodymium, promethium, and samarium, and heavy rare earth elements (HREE) - europium, gadolinium, terbium, dysprosium, holmium, erbium, thulium, ytterbium, lutetium, scandium, and yttrium.

How Are Rare Earths Used in Defense Applications?

It has been estimated that DOD uses less than 5% of domestic consumption of rare earths.[21] Rare earth elements are found in two types of commercially available, permanent magnet materials. They are samarium cobalt (SmCo), and neodymium iron boron (NdFeB). NdFeB magnets are considered the world's strongest permanent magnets and are essential to many military weapons systems. SmCo retains its magnetic strength at elevated temperatures and is ideal for military technologies such as precision-guided missiles, smart bombs, and aircraft. The superior strength of NdFeB allows for the use of smaller and lighter magnets in defense weapon systems.

The following illustrations (Figures 1-5) show the use of rare earth elements in a variety of defense-related applications:

- fin actuators in missile guidance and control systems, controlling the direction of the missile;
- disk drive motors installed in aircraft, tanks, missile systems, and command and control centers;
- lasers for enemy mine detection, interrogators, underwater mines, and countermeasures;
- satellite communications, radar, and sonar on submarines and surface ships; and
- optical equipment and speakers.

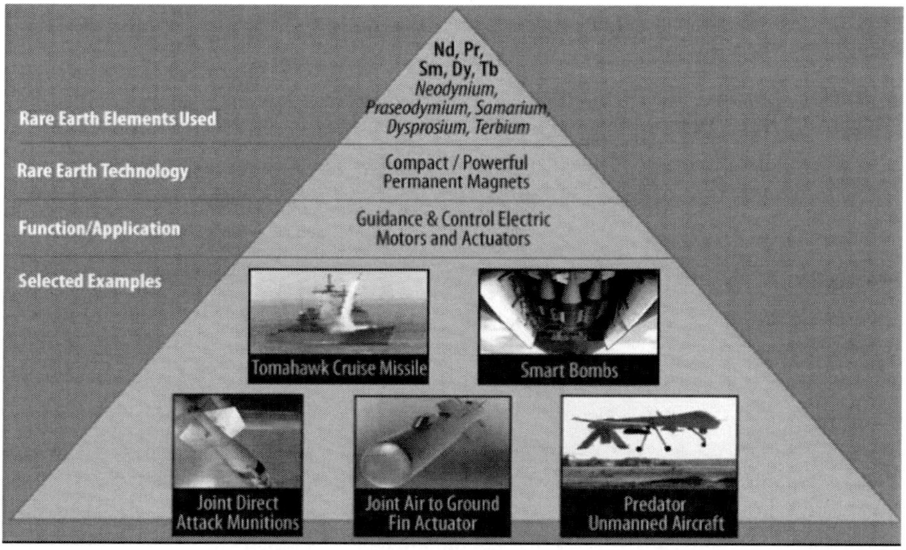

Source: Compiled from presentations by the Rare Earth Industry and Technology Association, the United States Magnet Manufacturing Association, and David Pineault, "Global Rare Earth Element Review," Defense National Stockpile Center, Spring 2010.

Figure 1. Rare Earth Elements in Guidance and Control Systems.

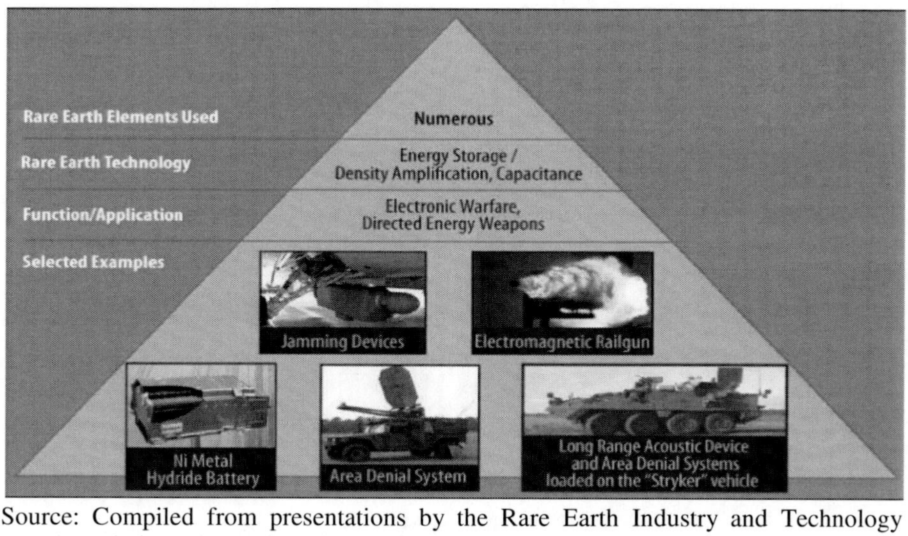

Source: Compiled from presentations by the Rare Earth Industry and Technology Association, the United States Magnet Manufacturing Association, and David Pineault, "Global Rare Earth Element Review," Defense National Stockpile Center, spring 2010.

Figure 2. Rare Earth Elements in Defense Electronic Warfare.

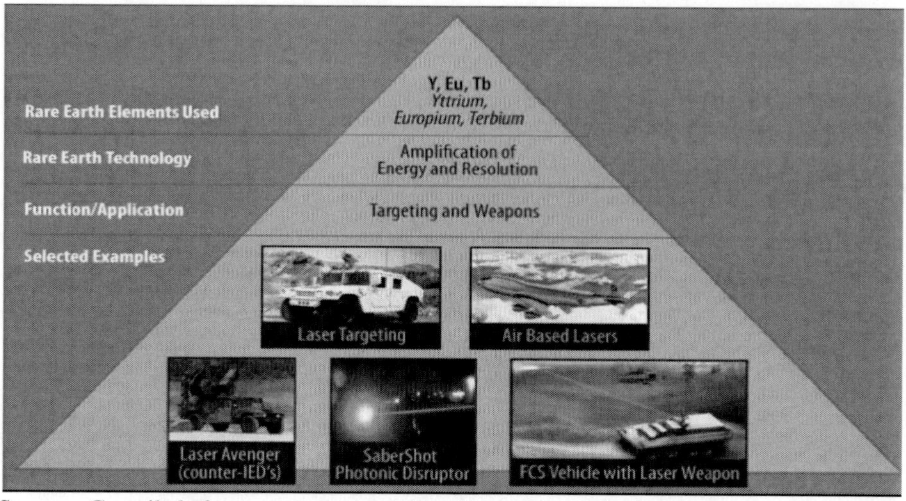

Source: Compiled from presentations by the Rare Earth Industry and Technology Association, the United States Magnet Manufacturing Association, and David Pineault, "Global Rare Earth Element Review," Defense National Stockpile Center, spring 2010.

Figure 3. Rare Earth Elements in Targeting and Weapon Systems.

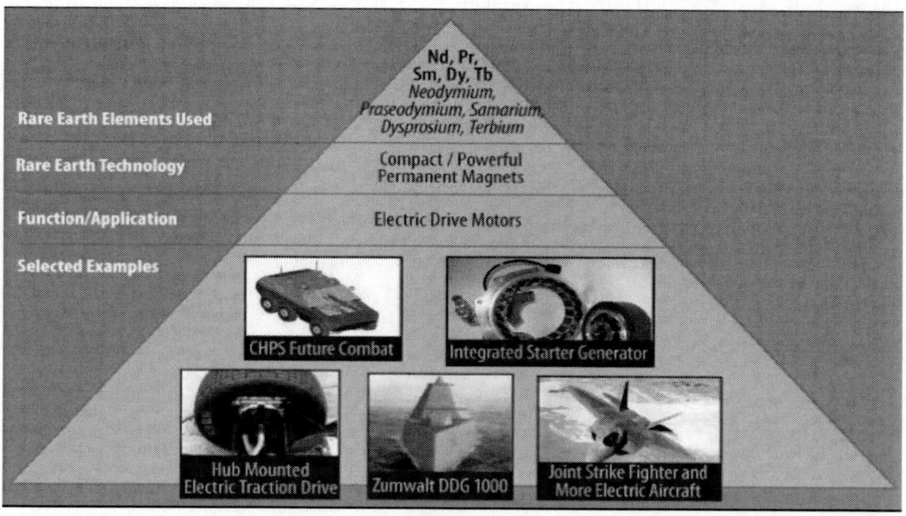

Source: Compiled from presentations by the Rare Earth Industry and Technology Association, the United States Magnet Manufacturing Association, and David Pineault, "Global Rare Earth Element Review," Defense National Stockpile Center, spring 2010.

Figure 4. Rare Earth Elements in Electric Motors.

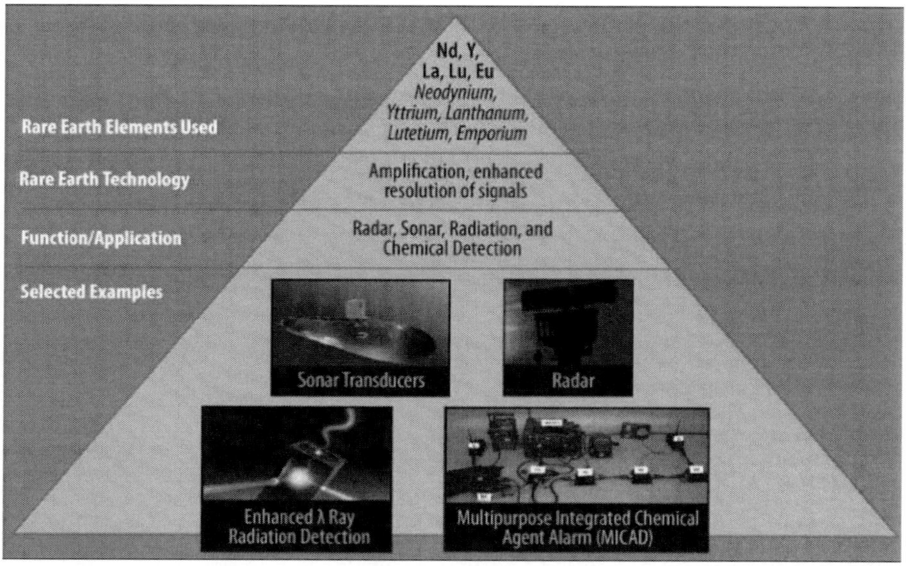

Source: Compiled from presentations by the Rare Earth Industry and Technology Association, the United States Magnet Manufacturing Association, and David Pineault, "Global Rare Earth Element Review," Defense National Stockpile Center, spring 2010.

Figure 5. Rare Earth Elements and Communication.

How and Where Are Rare Earths Produced?[22]

In April 2010, GAO reported on the world's production of rare earths and stated that China produced

- 97% of rare earth ore,
- 97% of rare earth oxides,
- 89% of rare earth alloys,
- 75% of neodymium iron boron magnets (NeFeB), and
- 60% of samarium cobalt magnets (SmCo).

The rare earth production process is complex and expensive. The stages of production consist of mining, separating, refining, alloying, and manufacturing rare earths into end-use items and components, as described in the GAO report.[23]

- The first stage is the actual mining, where the ore is taken out of the ground from the mineral deposits.
- The second stage is separating the ore into individual rare earth oxides.[24]
- The third stage is refining the rare earth oxides into metals with different purity levels; oxides can be dried, stored, and shipped for further processing into metals.
- The fourth stage is forming the metals, which can be processed into rare earth alloys.
- The fifth stage is manufacturing the alloys into devices and components, such as permanent magnets.

From the 1960s to the 1980s, the United States was the leader in global production of rare earths and in the research and development of high-performance magnets.[25] Since that time, as discussed above, production has shifted primarily to China, due to lower labor costs and lower environmental standards. China is the only exporter of commercial quantities of rare earth metals.[26]

Today, the United States almost entirely lacks the refining, fabricating, metal-making, alloying, and magnet manufacturing capacity to process rare earths. While the United States lacks domestic facilities that produce sintered NdFeB magnets, Molycorp does produce NeFeB and SmCo alloys which are used in permanent magnets.[27]

One U.S. company, Electron Energy Corporation (EEC) in Landisville, PA, produces SmCo permanent magnets. EEC, in its production of SmCo permanent magnets, uses predominately samarium metal and significant amounts of gadolinium, rare earths for which there is no U.S. production. Additional rare earth elements needed to produce rare earth magnets such as NeFeB include small amounts of dysprosium and possibly terbium. Currently, dysprosium and terbium are only available from China. EEC also imports metals for its magnet production from China through North American distributors and processes them into alloys in the United States before further processing into sintered SmCo magnets.[28]

Supply Chain Issues[29]

The supply chain for rare earth elements generally consists of mining, separation, refining, alloying, and manufacturing (devices and component parts). A major issue for REE development in the United States is the lack of refining, alloying, and fabricating capacity that could process any future rare earth production. One U.S. company, Electron Energy Corporation (EEC) in Landisville, PA, produces samarium cobalt (SmCo) permanent magnets, and Hitachi Metals, Ltd. of Japan is producing small amounts of the more desirable neodymium iron-boron (NdFeB) magnets (needed for numerous consumer electronics, energy, and defense applications) at its China Grove, North Carolina facility. EEC, in its production of its SmCo permanent magnet, uses small amounts of gadolinium—an REE of which there is no U.S. production. Even the REEs needed for these magnets that operate at the highest temperatures include small amounts of dysprosium and terbium, both available only from China at the moment. EEC imports magnet alloys used for its magnet production from China.

Prior to multimillion dollar investments in mining, separation, and alloying facilities by Molycorp, and other exploration and development projects in the United States, there was a significant underinvestment in U.S. supply chain capacity (including processing, workforce development, and research and development (R&D) which has left the United States nearly 100% import dependent on all aspects of the REE supply chain and dependent on a sole source for much of the material. An April 2010 GAO report illustrates the lack of U.S. presence in the REE global supply chain at each of the five stages - mining, separation, refining oxides into metal, fabrication of alloys and the manufacturing of magnets and other components. According to the GAO report, China produces about 95% of the REE raw materials, about 97% of rare earth oxides, and is the only exporter of commercial quantities of rare earth metals (Japan produces some metal for its own use for alloys and magnet production). About 90% of the metal alloys are produced in China (small production in the United States) and China manufactures 75% of the NeFeB magnets and 60% of the SmCo magnets. Thus, even if U.S. rare earth production ramps up, much of the processing/alloying and metal fabrication would occur in China.

According to investor analyst Jack Lifton, the rare earth metals are imported from China, then manufactured into military components in the United States or by an allied country. Lifton states that many investors believe that for financing purposes, it is not enough to develop REE mining operations alone without building the value-added refining, metal production, and alloying capacity that would be needed to manufacture component parts for end-use products. According to Lifton, vertically-integrated companies may be more desirable. It may be the only way to secure investor financing for REE production projects.[30] Joint ventures and consortiums could be formed to support production at various stages of the supply chain at optimal locations around the world. Each investor or producer could have equity and offtake commitments. Where U.S. firms and U.S. allies invest is important in meeting the goal of providing a secure and stable supply of REEs, intermediate products, and component parts needed for the assembly of end-use products.

Most experts have predicted where new mining capacity for rare earths is likely to come on-stream, but it is just as important to know where new downstream capacity (processing, refining, and metals alloying) is being built or likely to be built in the world as well as the likely investors in downstream capacity for rare earths. Additional questions that could be addressed by Congress include how long would it take to develop the skill set in the United States for downstream production activities? Would an international educational exchange program with those countries already involved in rare earth refining and recycling be appropriate?

Molycorp's "Mine to Magnet" Vertical Integration Approach for Rebuilding the U.S. Rare Earth Supply Chain

From the mid-1960s through the 1980s, Molycorp's Mountain Pass mine was the world's dominant source of rare earth oxides. The ramp up in production had been driven primarily by Molycorp's higher grade, its relatively low cost, and a rapid rise in the demand for the LREEs, particularly europium used for red phosphors in television and computer monitors, and cerium for glass polishing.[31] However, by 2000, nearly all of the separated rare earth oxides were imported, primarily from China. Because of China's oversupply, lower cost production, and a number of environmental (e.g., a pipeline spill carrying contaminated water) and regulatory issues at Mountain Pass, Molycorp ceased production at its mine in 2002. Since then, the United States has lost nearly all of its capacity in the rare earth supply chain, including intellectual capacity. However, under new ownership since 2008, Molycorp has embarked upon a campaign to change the rare earth position in the United States with its "mine to magnet" (vertical integration) business model.

After major energy producer Chevron purchased Union Oil Company of California (UNOCAL), which included the rare earth mine at Mountain Pass, Chevron wanted to focus on its energy business. They were willing to sell-off its non-energy Molycorp Mountain Pass asset.

When investor groups purchased Molycorp from Chevron in 2008, they did not inherit the environmental liability that resulted from the pipeline spill.

Chevron continued the cleanup that resulted from an earlier ruptured water disposal pipeline carrying some chemical contaminants from the oxide separation facility. Since its purchase by the new owners, Molycorp CEO and engineers have been adamant about minimizing their environmental footprint during the separation phase of the process.

Molycorp designed a proprietary oxide separation process that would use fewer reagents and recycle the waste water, thus doing without a disposal pond. Molycorp broke ground for their new separation facility at the Mountain Pass mine in 2011. This complex process separates out the individual elements which follows the mining of the raw material. Molycorp engineers suggest that they will use one-half the amount of ore to get the same amount of usable end product.[32] The chloralkali facility which would allow recycling of wastewater is now mechanically complete and is being commissioned.

Molycorp recently acquired the Japanese subsidiary Santoku America in Tolleson, AZ, and renamed it Molycorp Metals and Alloys (MMA). This acquisition is part of the firm's strategy to become a vertically integrated company. It produces both NdFeB and SmCo alloys used in the production of permanent magnets. Molycorp Metals and Alloys is the sole U.S. producer of the NdFeB alloy. Their intention is to modernize the facility and expand metals and metal alloy production.[33] Molycorp also recently purchased a 90.023% majority interest in AS Silmet, (renamed Molycorp Silmet) an Estonian-based rare earth element and rare metals processor, which will double its capacity for rare earth oxide and metal production (separation) in the near-term, according to Molycorp officials.

The management at MMA is also examining ways to improve metal recycling. Much of their recycling research is focused on the magnets and the highly valued HREEs. They want to probe into the commercial feasibility of recycling materials contained in permanent magnets used in consumer goods. Sourcing sufficient quantities of end-use materials and understanding the metallurgical processes for extracting the heavy rare earth elements such as the dysprosium and terbium is an important part of the research. Testing the quality of the recyclable material and evaluating the economics will determine the project's success. Molycorp is also evaluating near-term opportunities to recycle energy efficient light bulbs for the phosphors.[34]

Molycorp has entered into a cooperative research and development agreement (CRADA) with U.S. Department of Energy's Ames Laboratory to study new methods to create commercial-grade permanent magnets used in commercial applications. Development of downstream activities such as refining, rare earth metals alloying, and permanent magnet manufacturing will require a large amount of financing, a skilled workforce, and a sizeable U.S. market, all of which could be more completely developed in the long term. Keeping and recruiting top talent (in engineering, science, and finance) that can help Molycorp achieve its plans for vertical integration is one of the company's top priorities, according to company officials. Their aim is to consistently invest in the right people and the right training to accomplish its goal.

Recent supply chain developments by Molycorp include the acquisition of Neo Materials Technology, Inc.—a Toronto-based firm (renamed Molycorp Canada) with rare earth processing and permanent magnet powder facilities in China. According to Molycorp, 18% of Neo Materials production volume goes to domestic Chinese companies and 33% is directly exported to Japan plus an additional 11% goes to Japanese companies operating within China. One concern voiced by critics of this deal is that some of these highly valued materials could potentially become subject to Chinese export restrictions. Molycorp plans to ship some of its production capacity to its Neo Materials facility in China. Molycorp also entered a joint venture with Daido Steel and Mitsubishi Corporation of Japan and currently manufactures sintered permanent rare earth (NdFeB) magnets in Japan and sold on the world market.

Selected Recent Supply Chain Developments

Lynas and Siemens have entered into a joint venture for the manufacturing of magnets used in wind turbine generators. Lynas (45% stake) will provide raw material to Siemens (55% stake) from their Mt. Weld mine in Australia, which began production in 2012. Lynas is processing the concentrate at its Malaysian processing facility, which commenced operations in November 2012, after a long and contentious approval process with the Malaysian government. There are ongoing concerns in Malaysia over the proper disposal of thorium, which is contained in the mineral deposit and produced alongside the rare earth elements.

The Great Western Mineral Group (GWMG) will form a joint venture with China's Ganzhau Qiandong Rare Earth Group to build an oxide separation facility in South Africa. The raw material for the separation facility will be produced at GWMG's SKK mine in South Africa. A feasibility study of the project is underway.

Frontier Rare Earths, based in Luxemburg, along with Korea Resources Corp. formed a joint venture to build a separation facility also in South Africa. Frontier Rare Earths owns the nonproducing rare earth Zondkopsdrift mine in South Africa.

A Historical Perspective on the Potential Impact of Chinese Policies on Rare Earth Materials[35]

China's near monopoly of rare earth production and efforts over the past few years to restrict rare earth exports have raised concerns among U.S. policymakers.[36] For example, in July 2010, China announced that it would reduce its export quota of rare earth elements by 70% during the second half of 2010 over the previous year's level (or a 40% drop for the full year over 2009 levels).[37] In December 2010, China announced its export quota allocations for the first half of 2011, which, reportedly, were 35% less than the quota allocation level in the first half of 2010.[38] In July 2011, China announced that the quota level for the second half of 2011 (at 15,738 metric tons) would make the overall quota for 2011 roughly equal to 2010 levels. However, U.S. officials complained that China's second half 2011 quota levels included ferroalloys containing more than 10% rare earths, which had not been previously included in its rare earth quota levels, and hence, the new quota levels were more restrictive.[39] A U.S. Trade Representative (USTR) official was quoted as saying that the USTR continues to be "deeply troubled" by China's use of market distorting export restrictions on raw materials, including rare earths.[40] China's government responded by saying that it would apply the same policies to both domestic and overseas companies in rare earth production, processing, and export.[41] In addition to export quotas, China imposes export tariffs of 15% to 25% on rare earth as well.

In September 2010, China reportedly delayed shipments of rare earth to Japan (the world's largest rare earth importer) for about two months because of a territorial dispute. This has given rise to concerns that China may attempt use its control of rare earth as leverage to obtain its political and economic goals.[42]

The Chinese government has announced a number of initiatives over the past few years to further regulate the mining, processing, and exporting of rare earth elements, such as consolidating production among a few large state-owned enterprises and cracking down on illegal rare earth mining and exporting. The Chinese government contends that its goals are to better rationalize and manage its rare earth resources in order to slow their depletion, ensure adequate supplies and stable prices for domestic producers, obtain a more favorable return for exports, and reduce pollution. Critics of China's rare earth policies contend that they are largely aimed at inducing foreign high-technology and green technology firms to move their production facilities to China in order to ensure their access to rare earth elements, and to provide preferential treatment to Chinese high-tech and green energy companies in order to boost their global competitiveness. Such critics contend that China's restrictions of rare earth elements violate its obligations under the World Trade Organization (WTO).

Some have urged the USTR to bring a dispute resolution case against China in the WTO, similar to a case the United States brought against China in 2009 over its export restrictions (such as export quotas and taxes) on certain raw materials (including bauxite, coke, fluorspar, magnesium, manganese, silicon metal, silicon carbide, yellow phosphorus, and zinc). In that case, the United States charged that such policies are intended to lower prices for Chinese firms (especially the steel, aluminum, and chemical sectors) in order to help them obtain an unfair competitive advantage. China claims that these restraints are intended to conserve the environment and exhaustible natural resources.

In July 2011, a WTO panel issued a report that ruled that many of China's export restraints on raw materials violated WTO rules. In particular, the panel rejected China's argument that some of its export duties and quotas were justified because they related to the conservation of exhaustible natural resources for some of the raw materials. The panel stated that China was not able to demonstrate that it imposed these restrictions in conjunction with restrictions on domestic production or consumption of the raw materials so as to conserve the raw materials, protect the health of its citizens, or to reduce pollution.[43] China is appealing the WTO panel's ruling.

A March 2011 letter written by Senators Casey, Schumer, Stabenow, and Whitehouse urged the Obama Administration to instruct the U.S. executive director at each multilateral bank, including the World Bank, to oppose the approval of any new financing to the Chinese government for rare earth projects in China, including rare earth mining, smelting or separation, or production of rare earth products.[44] The letter also urged the Administration to impose the same types of restrictions on Chinese investment in mineral exploration and purchases in the United States as China imposes on foreign investment in rare earth in China.[45]

Some rare earth observers are concerned that Molycorp's purchase of Neo Materials Technologies could potentially make it more difficult for Molycorp to supply U.S. defense needs for rare earths, given China's domination and the increased broadening of restrictions on exporting rare earth minerals.[46]

Are Rare Earths Critical Materials for U.S. Defense?

DOD has not publicly stated whether rare earths fall within the context of materials considered strategic or critical for U.S. defense needs. There are several definitions of what constitutes a strategic or critical material; however, there is disagreement over which rare earth elements fall within these categories. Generally, strategic and critical materials have been associated with national security purposes. Some experts trace the first mention of strategic and critical materials to legislative language contained in both the Naval Appropriations Act of 1938 and the Strategic and Critical Materials Stockpiling Act of 1939 (P.L. 76-117, 50 U.S.C. 98 et seq.), which authorized the development of an inventory of strategic and critical materials for military use and provided funds for their purchase.[47]

DOD's current position on strategic materials was largely determined by the findings of the Strategic Materials Protection Board (SMPB).[48] The purpose of the SMPB was to determine the need to provide a long-term domestic supply of strategic materials designated as critical to national security, and to analyze the risk associated with each material and the effect on national defense that not having a domestic supply source might pose. The SMPB was to meet as determined to be necessary by the Secretary of Defense, but not less frequently than once every two years. SMPB's last report was issued in December 2008. Given the two-year meeting requirement, the board would have met in December 2010, but no meeting was held.

In the December 2008 report, the SMPB defined critical materials in this way: "the criticality of a material is a function of its importance in DOD applications, the extent to which DOD actions are required to shape and sustain the market, and the impact and likelihood of supply disruption."[49] Based on DOD's definition for "critical material," the 2008 SMPD report defined one material, beryllium, as a "strategic material critical to national security." The SPMB offered the following justification:

> High purity beryllium is essential for important defense systems, and it is unique in the function it performs. High purity beryllium possesses unique properties that make it indispensable in many of today's critical U.S. defense systems, including sensors, missiles and satellites, avionics, and nuclear weapons. The Department of Defense dominates the market for high purity beryllium and its active and full involvement is necessary to sustain and shape the strategic direction of the market. There is a significant risk of supply disruption. Without DOD involvement and

support, U.S. industry would not be able to provide the material for defense applications. There are no reliable foreign suppliers that could provide high purity beryllium to the Department. Recognizing that high purity beryllium meets all the conditions for being a critical material, the Department should take, and has taken, special action to maintain a domestic supply. The Department has used the authorities of Title III of the Defense Production Act to contract with U.S. firm Brush-Wellman, Inc. to build and operate a new high purity beryllium production plant.[50]

The House Armed Services Committee criticized this definition, as discussed in the following excerpt that appeared in the House report accompanying H.R. 2647, the FY2010 National Defense Authorization Act.[51]

This definition limits the purview of the Board to only those materials for which the determinations the Board is tasked to make are presupposed in the definition of the materials themselves. Furthermore, such a definition fails to include a range of materials that Congress has designated as critical to national security and, as such, has provided significant protection or domestic preference in DOD policy and in statute. For example, Congress has determined that reliance on foreign sources of supply for materials such as titanium, specialty steel, and high performance magnets, poses a heightened risk. The Board's narrowing of the definition of materials critical to national security renders the Board unable to provide perspective on the adequacy, suitability, or effectiveness of those policies. Moreover, it limits the ability of the Board to consider any course of action, however minor, in relation to a material until the point at which potential damage to national security is imminent and severe. It also creates the perverse situation that a material could be critical to every element of the industrial base upon which the Department depends, but not considered critical to the Department itself if the material is also used significantly in commercial items. As an indication of the inadequacy of this definition for the Board's functioning, the Board currently identifies only one material as meeting the definition for consideration as a strategic material critical to national security. The committee does not find this conclusion to be plausible and expects that the Board will swiftly revisit this definition to ensure that it is able to identify gaps in our domestic defense supply chain and provide the President, the ecretary of Defense, and Congress with information, analysis, and advice on strategic materials which are critical to the operations of the Department of Defense.

It should be noted that Congress has addressed this issue in the P.L. 111-383, FY2011 National Defense Authorization Act (H.R. 6523), where strategic materials are defined as "material essential for military equipment, unique in the function it performs, and for which there are no viable alternatives."

POLICY ISSUES FOR CONGRESS

Dependence on Foreign Sources for Rare Earth Materials

Some Members of Congress have expressed concern with the nearly total U.S. dependence on foreign sources for rare earth elements. Some have raised questions about China's near dominance of the rare earth industry and the implications for U.S. national security. Yet the "crisis" for many policymakers is not the fact that China has cut its rare earth exports and appears to be restricting the world's access to rare earths, but the fact that the United States has lost its domestic capacity to produce strategic and critical materials, and that the manufacturing supply chain for rare earths has largely migrated to outside the United States. Still others are concerned about the impact of a potential supply chain vulnerability of materials critical for defense systems. Additionally, some Members of Congress have questioned the lack of knowledge of what specific materials are needed for defense purposes, which materials are strategic and critical to national security, and what steps might be taken to increase the domestic capability to produce these materials.

In January 2011, three Members of Congress wrote a letter to Secretary of Defense Robert M. Gates outlining their concerns over what they perceived as a lack of action on DOD's part to ensure that adequate supplies of rare earths were available. They pressed for DOD to take immediate action, as described in excerpts below.

> Clearly, rare earth supply limitations present a serious vulnerability to our national security. Yet early indications are the DOD has dismissed the severity of the situation to date. Based on initial discussions with the DOD Office of Industrial Policy, we understand the effort to precisely ascertain and fully comprehend DOD consumption of certain rare earth elements is still an ongoing effort. In our view, it is a fundamental responsibility of DOD Industrial Policy to have a comprehensive understanding of the security of our defense supply chain, which requires

understanding detailed knowledge of the sources and types of components and materials found in our weapon systems.

As the ultimate customer, the Department has the right and responsibility to require their contractors to provide a detailed accounting of the various rare earth containing components within their weapon systems. This information should then be aggregated into an element by element overall demand for DOD. With that knowledge, DOD could compare expected supply and demand of each rare earth element with overall consumption by the Department to identify critical vulnerabilities in our supply chain. This will enable the Department to establish policies to ensure the defense supply chain has access to those materials. For example, one policy may be for the DOD to establish a limited stockpile of rare earth alloys that are in danger of supply interruption to ensure security of supply of both metals and magnets.[52]

Government Reports on Rare Earths

GAO Report on the Rare Earth Supply Chain

In response to congressionally directed requirements in Section 843 of the National Defense Authorization Act of 2010 (P.L. 111-84), GAO was asked to examine the defense rare earth supply chain issues. An April 2010 GAO report noted the lack of U.S. presence in the global supply chain at each of the five stages of rare earth production—mining, separating, refining (oxides into metal), fabricating (of alloys), and manufacturing (of magnets and other components). GAO concluded that the United States lacks a domestic rare earth supply chain and offered the following assessment of the current defense rare earth supply:

- While rare earth ore deposits are geographically diverse, current capabilities to process rare earth metals into finished materials are limited mostly to Chinese sources.
- The United States previously performed all stages of the rare earth material supply chain, but now most rare earth materials' processing is performed in China, giving it a dominant position that could affect worldwide supply and prices.
- Based on industry estimates, rebuilding a U.S. rare earth supply chain may take up to 15 years and is dependent on several factors, including securing capital investments in processing infrastructure, developing

new technologies, and acquiring patents, which are currently held by international companies.[53]

GAO was unable to determine whether DOD faces any supply chain vulnerability issues or the degree to which national security interests are potentially threatened by the current rare earth situation. Its assessment was limited, primarily because DOD stated that it was in the process of performing its own internal assessment and had not yet identified national security risks or taken steps to address any potential material shortages.

Annual Industrial Capabilities Report to Congress

Section 2504 of Title 10, United States Code, requires that the Secretary of Defense submit an annual report on industrial capabilities to the Committee on Armed Services of the Senate and the Committee on Armed Services of the House of Representatives. The 2009 report did not address the rare earth supply, but it did suggest that the issue warranted further study, as described in excerpts from the report.

> The lessons learned from the pre-slowdown economy will concentrate a global push for fuel efficiency and finding substitutes for hydrocarbon fuel products. This will drive up the demand for specialty metals and super alloys that are closely associated to battery manufacturing. These metals are typically not mined or melted within the United States and the E.U. countries. Therefore, this will likely become a growing strategic concern for the United States as resources will have to be utilized to secure the free flowing access to the limited supply of super alloys and specialty metals products (i.e., chromium, cobalt, lithium, rare earth and platinum group metals).[54]

U.S. Department of Energy Report on Critical Mineral Strategy

In December 2010, the U.S. Department of Energy released a report that examined the role of rare earths in renewable energy technologies. While the report did not focus on the use of rare earths for national security and defense purposes, it does shed light on the steps DOD has reportedly undertaken to review the rare earth supply chain, as described in excerpts below.

> Recognizing the evolution of the market for rare earth elements (REEs), in the summer of 2009 the Office of Industrial Policy/AT&L, Department of Defense (DOD) self-initiated a review of the U.S. supply

> chain. The study is based on available forecasts and data from multiple sources and as a result, most of the data are available only at the aggregate level of all REE [Rare earth elements]. The study reviews the U.S. supply chain for both commercial and defense demand of REE. The study also assesses gaps in the supply chain and their potential implications for the Department.
>
> The rationale for this effort included the U.S. dependence on a sole supplier that is not domestic, the importance of REE in certain defense applications and forecasts for a surge in demand for commercial end uses that could strain global supplies. Recent events in the global market for REE have reinforced the Department's concern regarding reliable and secure supplies of REE.[55]

Coordination of the Federal Approach to Rare Earths

Rare earths are used for both commercial and defense purposes, but the majority of domestic use of rare earths is for commercial purposes. There is no unified opinion on whether every rare earth element is considered "critical" or "strategic" for economic or national security purposes, or whether economic security is a national security issue.

Working with the Departments of Commerce and Energy, the White House Office of Science and Technology Policy (OSTP) began gathering experts to hold interagency group discussions on rare earth elements from 2007 to 2008. Initially, an interagency working group (an ad hoc working group) was the result of a roundtable discussion on rare earths organized by the Department of Commerce. This group of subject-matter experts from various federal agencies discussed the potential usefulness of the White House taking a lead role on rare earth strategy.[56]

Absence of the Study of Rare Earth Application Sciences in Most U.S. Colleges and Universities

There is a growing gap between the United States and China with regard to the academic study of rare earth elements. China employs thousands of scientists in both disciplines. The only U.S. public university with a rare earths specialty is the Colorado School of Mines (MINES); a public research university devoted to engineering and applied science. MINES is one of a few academic resources in the world that provides broad experience in mineralogy, resource exploration, mining, extraction, and production.

In a hearing before the House Committee on Science and Technology, Subcommittee on Investigations and Oversight, Dr. Stephan Freiman, a scientist and former member of the National Research Council's (NRC) Committee on Critical Mineral Impacts on the U.S. Economy, discussed the conclusions of a study sponsored by the NRC to examine the role of nonfuel minerals in the U.S. economy and potential material supply vulnerabilities.[57] Among the study's recommendations were the following:

> Federal agencies, including the National Science Foundation, Department of the Interior (including the USGS), Department of Defense, Department of Energy, and Department of Commerce, should develop and fund activities, including basic science and policy research, to encourage U.S. innovation in the area of critical minerals and materials and to enhance understanding of global mineral availability and use programs involving academic organizations, industry, and government to enhance education and applied research.

The study also recommended funding scientific research on the entire mineral life cycle and building cooperative programs among academia, industry, and government to enhance education and applied research.[58]

Options for Congress

Congress may consider both short-range and long-range options for securing a source for rare earth elements as part of its oversight role in addressing U.S. national security interests. Short-range options potentially include hearings on the "Section 843" rare earths report, convening defense suppliers to discuss rare earth material shortages, establishing rare earth material stockpiles for defense purposes, instituting a new critical minerals program, and reconvening the SMPB. Long-range options could include reducing DOD consumption of rare earth elements by identifying and securing equally effective alternatives to rare earths, establishing partnerships with foreign allies that could potentially offer a diversified source of foreign suppliers outside of China, and providing more financial assistance for rare earth production within the United States. Each of these potential options is discussed below.

Congressional insight on these potential actions will largely depend on the findings and conclusions reached in DOD's long-overdue self-assessment on the defense rare earth supply chain. However, it is not clear if or when DOD will release its report.

Hearings on the DOD Report on the Assessment of the Rare Earth Materials Supply Chain

Congress could require DOD to release the report, and then hold hearings to examine DOD's assessment and conclusions. The report was required to be released within 180 days of the enactment of the act, which would have been on or about July 7, 2011. The reasons for the delay are uncertain.

Convene Defense Suppliers to Discuss Supply Chain Issues

Congress could meet with defense suppliers, at all tiers of the supply chain, to ascertain their knowledge of material shortages and bottlenecks. While DOD purchases the end product (the weapons system) from prime contractors and relies on prime contractors to deliver the finished product, rare earth elements are important throughout the supply chain from the prime contractor through successive subcontractor tiers. Some contractors at lower ends of the tiers may be reluctant to signal to DOD that there are supply chain issues or challenges.

An issue that warrants further understanding is where there is convergence between the rare earth value supply chain and the defense supply chain. The rare earth supply chain starts with mining, flows from ore to concentrate, to oxide, to metal, to alloy, and then to the finished product, the magnet. In contrast, the defense supply chain starts with the prime contractor and moves through a successive number of subcontractors down to the ultimate "first line processor" who purchases a rare earth, value-added product such as metal, alloy, or permanent magnets for incorporation into a defense component.

Convene the Strategic Materials Protection Board

Congress could require DOD to convene the Strategic Materials Protection Board (SMPB). In its December 2008 report, as discussed above, the SMPB defined critical materials in this way: "the criticality of a material is a function of its importance in DOD applications, the extent to which DOD actions are required to shape and sustain the market, and the impact and likelihood of supply disruption."[59] As a result, the SMPB defined only one rare earth element, beryllium, as a "strategic material critical to national security." Congress may convene the board because the present board might determine that some rare earth elements have moved into a position where they are now more critical to national security purposes. The next SMPB might determine that some rare earth elements have moved into a position where they are now more critical to national security.

Congress might demand the 2010 statutorily required meeting of the board to commence immediately. The next SMPB will be required to use the new definition of "materials critical to national security" as defined in Section 829 of the FY2011 NDAA, which states the following:

> Sec. 829. Definition of Materials Critical To National Security
>
> (1) The term "materials critical to national security" means materials (A) upon which the production or sustainment of military equipment is dependent; and (B) the supply of which could be restricted by actions or events outside the control of the Government of the United States.[60]

In the short run, however, creating a stockpile could raise prices even further because of the increased demand.

Require Stockpiling of Specific Materials

Congress could require a strategic rare earth elements stockpile. Stockpiles might possibly increase the security of the domestic U.S. supply for rare earths. Congress may consider compiling a "virtual" stockpile database, with commitments and contracts with suppliers to buy the items when needed. One trade association, USMMA, advocates for a limited strategic reserve of rare earth alloys, metals, and magnets. USMMA asserts that government action is needed to ensure that there is a downstream domestic manufacturing capability.

> This strategic stockpile would ensure our Department of Defense has ready access to those materials needed to ensure our national security and to incentivize the return of domestic manufacturing. With defense critical materials such as dysprosium being sourced solely from China, it is critical that the Department of Defense have access to rare earth oxides from reliable producers and manufacturers in the United States and ally nations to perform value added processes, such as metal, alloy and magnet manufacturing.[61]

Fund the Downstream Supply Capacity

Once DOD and its suppliers identify whether and where material shortfalls exist, Congress could determine which stages of the supply chain (e.g., mining or manufacturing) require federal funding.

Fund Rare Earth Research

With the growing strategic importance of rare earths, and in order to create interest and build additional U.S. leadership in rare earth research and development, Congress may consider funding rare earth application sciences in curriculums for military and other government institutes or in national research and development centers designed to train students, scientists, and engineers.

Institute a New Critical Minerals Program

Should DOD determine that rare earths fall into the classification of critical minerals, Congress could institute a new Critical Minerals Program. In the early 1980s, there existed a Critical Minerals Program aimed at warning Congress about potential supply shortages, protecting strategic materials, and keeping an inventory of those minerals on hand in order to mitigate a supply shock.[62] This program ended in the 1990s as the consensus within Congress grew that the market could handle mineral supply disruptions without government intervention. Two decades later, at a 2010 hearing of the House Science and Technology Committee on rare earths, one policymaker suggested that the time has come to revive the program:

This is not the first time the Committee has been concerned with the competitive implications of materials such as rare earths. In 1980—30 years ago—this Committee established a national minerals and materials policy. One core element in that legislation was the call to support for "a vigorous, comprehensive and coordinated program of materials research and development."

Unfortunately, over successive administrations, the effort to keep that program going fell apart. Now, it is time to ask whether we need to revive a coordinated effort to level the playing field in rare earths.

In particular, I want to learn if there is a need for increased research and development to help address this Nation's rare earth shortage, or if we need to re-orient the research we already have underway.

Based on my review of the written submissions, it appears that we could benefit from more research both in basic and applied materials sciences.[63]

Develop Partnerships with Allies to Diversify the Supply Source

Congress may encourage DOD to pursue joint ventures with other nations, as many other nations are seeking alternatives to a near total dependence on rare earths from China. These partnerships may take place at any stage of the supply chain. It is critical for DOD to consider the implications of sourcing used by these partner nations. For example, if DOD relies on a partner nation for rare earth metals, and that nation procures their oxides from China, this partnership may not provide the requisite security of supply.

APPENDIX. SELECTED LEGISLATIVE ACTIVITY

Some Members of Congress have introduced rare-earth related bills during the 112th and 111th Congresses. Some selected measures, related to national defense issues, are described below.

Legislation Introduced in the 112th Congress

P.L. 112-239 (H.R. 3310, 112th Congress), the National Defense Authorization Act for FY2013

H.R. 4310 was introduced on March 29, 2012, passed the House on May 18, 2012, and referred to the Senate Armed Services Committee on June 19, 2012. The bill has several rare earth-related provisions, as described below.

Section 901. Additional Duties of Deputy Assistant Secretary of Defense for Manufacturing and Industrial Base Policy and Amendments to the Strategic Materials Protection.

- Appoints Assistant Deputy Secretary of Defense for Manufacturing and Industrial Base Policy as Chair of the Strategic Materials Protection Board Chair;
- Requires that Board's findings be reviewed by the Secretary of Defense, congressional defense committees of Congress, and published in the Federal Register within 90 days of the Board meeting;
- Broadens the scope of duties assigned to the Office of the Assistant Deputy Secretary of Defense for Manufacturing and Industrial Base Policy to better provide oversight of the defense supply chain, including contractors and strategic materials, to ensure that there are no supply chain vulnerabilities regarding the nation's defense requirements;
- Specifically, "ensuring reliable sources of materials critical to national security, such as specialty metals, armor plate, and rare earth elements."[64]

Section 1603. National Security Strategy for National Technology and the Industrial Base.

- Requires the Secretary of Defense to develop a national security strategy for the national technology and industrial base, based on a prioritized assessment of risks and challenges to the defense supply chain, and ensuring that the national technology and industrial base is capable of achieving certain national security objectives;
- Ensuring reliable sources of materials critical to national security, such as specialty metals, armor plates and rare earth elements;

- Reducing both the presence and risk of counterfeit parts in the supply chain.[65]

S. 3254, the proposed National Defense Authorization Act for FY2013, was introduced on June 4, 2012, and referred to the Senate Armed Services Committee. While the enacted bill[66] did not include a provision on rare earths, Senate Rept. 112-173 contained a statement that reflected the Committee's view on the role of rare earth materials for defense purposes, as repeated here.[67]

> ***Essential role of rare earth materials***
>
> Rare earth materials play an essential role in several critical weapons components and systems such as precision-guided munitions, electric ship drives, command and control centers, and aircraft, tanks, and missile systems. The committee notes the predominance of unreliable foreign sources for rare earth materials, including China, which provides roughly 94 percent of the world's rare earth oxides and nearly all rare earth metal within the defense-related supply chain and which has repeatedly decreased export quotas and imposed embargoes of these critical materials. Even with the development of the domestic-supply chain there may be continued reliance on production of certain heavy rare earth elements from China. The importance of rare earth materials for national defense applications necessitates a thorough understanding of vulnerabilities in the rare earth supply chain and the development of pragmatic, actionable risk mitigation plans to reduce the likelihood of supply interruptions. The committee encourages the Department of Defense to carefully consider the role of U.S. producers and potential means to develop reliable domestic sources to meet Department rare earth materials requirements.[68]

P.L. 111-393, the Ike Skelton National Defense Authorization Act for FY2011

Section 843 of the Ike Skelton National Defense Authorization Act for FY2011 (P.L. 111-383) and S.Rept. 111-201 (accompanying S. 3454, the proposed Senate National Defense Authorization Act for FY2011) required the Secretary of Defense to conduct an assessment of rare earth supply chain issues and develop a plan to address any vulnerabilities.

Section 843

Section 843 required the Secretary of Defense, within 180 days of enactment of the act,[69] to report to Congress with an assessment of the supply and demand for rare earth materials in defense applications. The assessment would identify whether any rare earth materials would be:

> (1) Critical to the production, sustainment, or operation of significant United States military equipment; or
> (2) Subject to interruption of supply, based on actions or events outside the control of the Government of the United States.[70]

For every rare earth material identified that would meet these criteria, the Secretary of Defense would develop a plan for the long-term availability of such materials with the goal of establishing "an assured source of supply of such material in critical defense applications by December 31, 2015." The plan would consider the following:

> (1) An assessment of whether the material should be included in the National Defense Stockpile;
> (2) in consultation with the United States Trade Representative, the identification of any trade practices known to the Secretary that limit the Secretary's ability to ensure the longterm availability of such material or the ability to meet the goal of establishing an assured source of supply of such material by December 31, 2015;
> (3) An assessment of the availability of financing to industry, academic institutions, or not-for-profit entities to provide the capacity required to ensure the availability of the material, as well as potential mechanisms to increase the availability of such financing;
> (4) An assessment of the benefits, if any, of Defense Production Act funding to support the establishment of an assured source of supply for military components;
> (5) An assessment of funding for research and development
> (6) Any other risk mitigation method determined appropriate by the Secretary that is consistent with the goal of establishing an assured source of supply by December 31, 2015; and,
> (7) For steps of the rare earth material supply chain for which no other risk mitigation method, as described in paragraphs (1) through (6), will ensure an assured source of supply by December 31, 2015, a specific

plan to eliminate supply chain vulnerability by the earliest date practicable.

H.R. 4402, the National Strategic and Critical Minerals Production Act

H.R. 4402, the National Strategic and Critical Minerals Production Act of 2012, was introduced on Apri19, 2012, passed the House on July 12, 2012, and was referred to the Senate on July 16, 2012. The bill would have required both the Secretary of the Interior and the Secretary of Agriculture to more efficiently develop domestic sources of the minerals and materials of strategic and critical importance to U.S. economic and national security, and manufacturing competitiveness.[71]

H.R. 3449, Defense Supply Chain and Industrial Base Security Act

H.R. 3449 was introduced by Representative Paul Ryan on November 16, 2011, and referred to the House Armed Services Committee. The bill would have required the Secretary of Defense to develop a defense supply chain and industrial base strategy, and subsequent plan, designed to secure the supply chain and industrial base sectors that the Secretary judges critical to U.S. national security.

H.R. 2184, Rare Earth Policy Task Force and Materials Act

H.R. 2184 was introduced by Representative Mike Coffman on June 15, 2011. The bill would establish a Rare Earth Policy Task Force for a period of 10 years within the Department of the Interior for the purpose of developing a plan to ensure the long-term supply of rare earth materials.[72] The Task Force would be directed to assist federal agencies in reviewing laws, regulations, and policies that discourage investment in, exploration for, and development of, domestic rare earths. The Task Force would also be required to submit an annual report that would provide a plan for research, development, demonstration, and commercial application to ensure the long-term, secure, and sustainable supply of rare earth materials sufficient to satisfy the national security, economic well-being, and industrial production needs of the United States (based on specific criteria). The bill was referred to the Committee on Natural Resources and to the Committee on Science, Space, and Technology.

H.R. 2090, Energy Critical Elements Advancement Act of 2011

H.R. 2090 was introduced by Representative Randy Hultgren on June 2, 2011, and was referred to the Subcommittee on Energy and the Environment. The bill would require the Secretaries of Energy and Interior to establish a research program to advance basic materials science, chemistry, physics, and engineering associated with energy critical elements.

H.R. 2011, National Strategic and Critical Minerals Policy Act of 2011

H.R. 2011 was introduced on May 26, 2011, by Representative Doug Lamborn and referred to the House Committee on Natural Resources, Subcommittee on Energy and Mineral Resources. On July 20, 2011, the committee ordered the bill to be reported, as amended. On October 14, 2011, the bill was amended by the Committee on Natural Resources and placed on the Union Calendar.

The bill would require the Secretary of the Interior to (1) conduct an assessment of the United States' capability to meet current and future demands for the minerals critical to domestic manufacturing competitiveness, economic, and national security in a time of expanding resource nationalism; (2) conduct an assessment of the current mineral potential of federal lands, and an evaluation of mineral requirements to meet current and emerging needs for economic and national security, and U.S. industrial manufacturing needs (such an assessment would address the implications of any potential mineral shortages or supply disruptions, as well as the potential impact of U.S. dependence on foreign sources for any minerals); and (3) conduct an inventory of rare earth elements and other minerals deemed critical based on the potential for supply disruptions, including an analysis of the supply chain for each mineral.

Finally, the bill would set policy goals for federal agencies to coordinate responsibilities for:

- facilitating the availability, development, and production of domestic mineral resources to meet national needs;
- promoting the development of economically sound and stable policies for domestic industries that promote mining, materials, and metals processing;
- creating a mechanism for assessing the U.S. mineral demand, supply, and needs; and
- minimizing duplication and delays in administering federal and state laws, regulations, and permit issuance and authorizations necessary to explore, develop, and produce minerals, and build and operate mineral-related facilities.[73]

H.R. 2284, Responsible Electronic Recycling Act

H.R. 2284 was introduced by Representative Gene Green on June 22, 2011, and on June 29 was referred to the Subcommittee on Energy and the Environment. This bill would include the establishment of a Rare Earth Materials Recycling Initiative, designed to assist in and coordinate the development of research in the recycling of rare earth materials found in electronic devices.

H.R. 1875, Building Our Clean Energy Future Now Act of 2011

H.R. 1875 was introduced by Representative David Cicilline on May 12, 2011, and referred to the House Committees on Ways and Means, Transportation and Infrastructure, Energy and Commerce, and Science, Space, and Technology. On May 26, 2011, the bill was referred to the Subcommittee on Energy and Environment. The bill seeks to lower gas prices by making investments in cleaner energy technologies and infrastructure.

H.R. 1388, the Rare Earths Supply Chain Technology and Resources Transformation Act of 2011

H.R. 1388 was introduced by Representative Mike Coffman on May 6, 2011, and referred to the House Committee on Science, Space, and Technology, Subcommittee on Energy and the Environment, and the Committees on Natural Resources and Armed Services. The bill is also referred to as the Restart Act of 2011. The bill seeks to reestablish a competitive domestic rare earths supply chain within DOD's Defense Logistics Agency (DLA).

H.R. 1540, the National Defense Authorization Act for FY2012

H.R. 1540 was introduced by Representative Howard McKeon on April 14, 2011. Section 835 would require the DLA Administrator for Strategic Materials to develop an inventory of rare earth materials to support defense requirements, as identified by the report required by Section 843 of the Ike Skelton National Defense Authorization Act for FY2011 (P.L. 111-383). Also, Amendment #87 to H.R. 1540 would require the Secretary of Defense to report back to Congress on the feasibility and desirability of recycling, recovering, and reprocessing rare earth elements, including fluorescent lighting used in DOD facilities. H.R. 1540 (H.Rept. 112-78) passed the House, May 26, 2011. On June 6, 2011, it was received in the Senate and referred to the Senate Committee on Armed Services.

S. 734, the Advanced Vehicle Technology Act of 2011

S. 734 was introduced by Senator Debbie Stabenow on April 5, 2011, and referred to the Committee on Natural Resources. The proposed bill would create a basic and applied research program, within the Department of Energy (DOE), focused on the development and engineering of new vehicle technologies. DOE is to promote, among many other goals, the exploration of substitutes and recycling of potential critical materials, including rare earth elements and precious metals. The Senate Committee on Energy and Natural Resources held a hearing on May 19, 2011.

H.R. 1367, the Advanced Vehicle Technology Act of 2011

H.R. 1367 was introduced by Representative Gary Peters on April 5, 2011, and referred to the Committee on Science, Space and Technology. On April 7, 2011, the bill was referred to the Subcommittee on Energy and Environment. S. 734 and H.R. 1367 are similar.

H.R. 1314, the Resource Assessment of Rare Earths (RARE) Act of 2011

H.R. 1314 was introduced by Representative Henry Johnson on April 1, 2011, and on April 6 was referred to the House Natural Resources Committee, Subcommittee on Energy and Mineral Resources. The bill would direct the Secretary of the Interior, through the Director of the U.S. Geological Survey, to examine the need for future geological research on rare earth elements and other minerals and determine the criticality and impact of a potential supply restriction or vulnerability.

H.R. 952, the Energy Critical Elements Renewal Act of 2011

On March 8, 2011, Representative Brad Miller introduced the Energy Critical Elements Renewal Act of 2011. The bill was referred to the Committee on Science, Space, and Technology. The bill would develop an energy critical elements program, amend the National Materials and Minerals Policy Research and Development Act of 1980, establish a temporary program for rare earth material revitalization, and serve other purposes.

S. 383, the Critical Minerals and Materials Promotion Act of 2011

On February 17, 2011, Senator Mark Udall introduced the Critical Minerals and Materials Promotion Act of 2011. One June 9, 2011, the bill was referred to the Committee on Energy and Natural Resources, Subcommittee on Energy. The bill was referred to the Committee on Energy and Natural Resources. The bill would require the Secretary of the Interior to establish a scientific research and analysis program to assess current and future critical mineral and materials supply chains, strengthen the domestic critical minerals and materials supply chain for clean energy technologies, strengthen education and training in mineral and material science and engineering for critical minerals and materials production, and establish a domestic policy to promote

> an adequate and stable supply of critical minerals and materials necessary to maintain national security, economic well-being, and industrial production with appropriate attention to a long-term balance between resource production, energy use, a healthy environment, natural resources conservation, and social needs.[74]

H.R. 618, the Rare Earths and Critical Materials Revitalization Act of 2011

On February 10, 2011, Representative Leonard Boswell introduced the Rare Earths and Critical Materials Revitalization Act of 2011. The bill was referred to the Committee on Science, Space, and Technology.

The bill seeks to develop a rare earth materials program and amend the National Materials and Minerals Policy, Research and Development Act of 1980. If enacted, it would provide for loan guarantees to revitalize domestic production of rare earths in the United States.

S. 1113, the Critical Minerals Policy Act of 2011

On May 26, 2011, Senator Lisa Murkowski introduced the Critical Minerals Policy Act of 2011, which was referred to the Committee on Energy and Natural Resources. On June 9, 2011, the Subcommittee on Energy held a hearing. The bill generally defines what critical minerals are but would request that the Secretary of the Interior establish a methodology (in consultation with others) that would identify which minerals qualify as critical. The Secretary of the Interior would direct a comprehensive resource assessment of critical mineral potential in the United States, including details on the critical mineral potential on federal lands. S. 1113 would establish a Critical Minerals Working Group to examine the permitting process for mineral development in the United States and facilitate a more efficient process, specifically, a draft performance metric for permitting mineral development and report on the timeline of each phase of the process. The DOI would produce an Annual Critical Minerals Outlook report that would provide forecasts of domestic supply, demand, and price for up to 10 years. DOE would lead research and development on critical minerals and workforce development that would support a fully integrated supply chain in the United States. Title II of the bill recommends mineral-specific action (led by DOE) for cobalt, helium, lead, lithium, low-btu gas, phosphate, potash rare earth elements, and thorium. Title III would, among other things, authorize for appropriation $106 million.

Legislation Introduced in the 111th Congress

In the 111th Congress, two bills were enacted that contain provisions affecting rare earth policy. The first was P.L. 111-84 (H.R. 2647), the National Defense Authorization Act for FY2010. Section 843 of P.L. 111-84 required GAO to examine rare earths in the defense supply chain, and it also required the Secretary of Defense to assess the defense supply chain and develop a plan to address any shortfalls or other supply chain vulnerabilities.[75] The second bill was P.L. 111-383, the Ike Skelton National Defense Authorization Act for FY2011, which contains a provision (Section 843) that requires the Secretary of Defense to undertake an assessment of the supply chain for rare earth materials and determine which, if any, rare earths are strategic or critical to national security and to develop a plan to address any supply chain vulnerabilities.[76] Other legislative provisions are listed below.

H.R. 4866, the Rare Earths Supply-Chain Technology and Resources Transformation Act of 2010

On March 17, 2010, Representative Mike Coffman introduced the Rare Earths Supply-Chain Technology and Resources Transformation Act of 2010 (RESTART). The bill was referred to three committees: the House Armed Services Committee, the House Ways and Means Subcommittee on Trade, and the House Financial Services Committee.

The bill sought to create a new interagency initiative on rare earth supply chain issues. H.R. 4866 would have established a federal government-wide interagency working group, at the Assistant Secretary level, from the Departments of Commerce, Defense, Energy, the Interior, and State, with participants from the U.S. Trade Representative (USTR) and White House Office of Science and Technology Policy. The working group would have assessed the rare earth supply chain to determine which rare earths were critical to national and economic security. Based on a critical designation, rare earth elements would have been stockpiled by the Defense Logistics Agency (DLA) as part of the National Defense Stockpile. The DLA would have made, if necessary, a commitment to purchase rare earth raw materials for processing and refining, including purchases from China. Stockpiling would have been terminated when the working group agencies determined that rare earths were no longer critical to U.S. national security or economic wellbeing.[77]

H.R. 6160, the Rare Earths and Critical Materials Revitalization Act of 2010

On September 22, 2010, Representative Kathleen Dahlkemper introduced the Rare Earths and Critical Materials Revitalization Act of 2010. The bill sought to develop a rare earth materials program and amend the National Materials and Minerals Policy, Research and Development Act of 1980. If enacted, the bill would have provided for loan guarantees to revitalize domestic production of rare earths in the United States. The bill was passed by the House on September 29, 2010, and forwarded to the Senate Committee on Energy and Natural Resources.

S. 3521, the Rare Earth Supply Technology and Resources Transformation Act of 2010

S. 3521 was introduced by Senator Lisa Murkowski on June 22, 2010. Congress held a hearing on the bill before the Senate Committee on Energy and Natural Resources, Subcommittee on Energy, on September 30, 2010. The text of the bill offered a "Sense of the Congress" statement that

> (1) the United States faces a shortage of key rare earth materials that form the backbone of both the defense and energy supply chains; (2) the urgent need to reestablish a domestic rare earth supply chain warrants a statutory prioritization of projects to support such reestablishment; (3) there is a pressing need to support innovation, training, and workforce development in the domestic rare earth supply chain; and (4) the Departments of Energy, of the Interior, of Commerce, and of Defense should each provide funds to academic institutions, federal laboratories, and private entities for innovation, training, and workforce development in the domestic rare earth supply chain.

S. 4031, the Rare Earths Supply-Chain Technology and Resources Transformation Act of 2010

S. 4031 was introduced by then-Senator Evan Bayh on December 15, 2010, and referred to the Senate Committee on Energy and Natural Resources. The bill would have promoted exploration and development of a domestic supply of rare earths, and reestablished a U.S. competitive rare earth supply chain for rare earths in the United States and in the countries of foreign allies.

End Notes

[1] Long, Keith R., S. Van Gosen, Bradley, Foley, Nora K., and Cordier, Daniel. The Principal Rare Earth Elements Deposits of the United States—A Summary of Domestic Deposits and a Global Perspective. *USGS Scientific Investigations Report, 2010-5220*, November 16, 2020, 96 p., at http://pubs.usgs.gov/sir/2010/5220/. Also see USGS Fact Sheet 087-02, and USGS Mineral Commodity Summaries, January 2011, at http://minerals.usgs.gov/minerals/pubs/commodity/rare_earths/mcs-2011-raree.pdf.

[2] According to then Andy Davis, Manager of Public Affairs for Molycorp, the company was stockpiling heavy concentrates known as SEG (primarily consisting of samarium, europium, and gadolinium.)

[3] H.Rept. 113-102, to accompany H.R. 1960, the proposed National Defense Authorization Act for Fiscal Year 2014.

[4] H.Rept. 113-102, to accompany H.R. 1960, the proposed National Defense Authorization Act for Fiscal Year 2014.

[5] S. 1600, the Critical Minerals Policy Act of 2013.

[6] H.R. 4402, the National Strategic and Critical Minerals Production Act of 2012 was introduced on July 16, 2012, and referred to the Committee on Energy and Natural Resources.

[7] For further discussion on the WTO dispute resolution case against China, see CRS Report R42510, *China's Rare Earth Industry and Export Regime: Economic and Trade Implications for the United States*, by Wayne M. Morrison and Rachel Y. Tang.

[8] Wan, William, Richburg, Keith B., and Nakamura, David. U.S. Challenges China's Curbs on Mineral Exports; China Vows to Strike Back. *Washington Post*, updated March 13, 2012; and Obama Announces WTO Case Against China Over Rare Earths. *CNN Online*, March 13, 2012, http://www.cnn.com/2012/03/13/world/asia/china-rare-earths-case/ index.html.

[9] China Sets Up Rare Earth Body to Shake Up Industry. Scientific American, April 10, 2012, http://www.reuters.com/ article/2012/04/10/us-china-rareearth-idUSBRE83707G20120410.

[10] Office of the U.S. Trade Representative. United States Challenges China's Export Restraints on Rare Earths. Statement of Ron Kirk, U.S. Trade Representative, March 13, 2012, http://www.ustr.gov/about-us/press-office/pressreleases/2012/march/united-states-challenges-china's-export-restraints-r.

[11] Friedman, Jennifer. WTO To Investigate China Curbs on Rare Earth Exports. *Bloomberg.com*, July 24, 2012. Accessed online at http://www.bloomberg.com/news/2012-07-23/wto-to-investigate-chinese-curbs-on-rare-earthexports.html. Also, see Gillespie, Clara and Pfeiffer, Stephanie. The Debate over Rare Earths: Recent Developments in the WTO Case. *The National Bureau of Asian Research*, July 11, 2012. Accessed online at http://www.nbr.org/research/activity.aspx?id=261. Also, see http://www.ustr.gov/trade-topics/enforcement/dispute-settlement-proceedings/wto-dispute-settlement/pending-wto-disputes-1.

[12] U.S. Department of Defense. Office of the Deputy Assistant Secretary of Defense for Manufacturing and Industrial Base Policy. Annual Industrial Capabilities Report to Congress, October 2013, p. 25. Accessed online at http://www.acq.osd.mil/mibp/docs/annual_ind_cap_rpt_to_congress-2013.pdf.

[13] Office of the Under Secretary of Defense for Acquisition, Technology and Logistics. Strategic and Critical Materials 2013 Report on Stockpile Requirements. January 2013. Accessed online at http://mineralsmakelife.org/assets/images/content/resources/Strategic_and_Critical_Materials_2013_Report_on_Stockpile_Requirements.pdf.

[14] Ibid., pages 5-6.

[15] Ibid., Figure 3. Six Rare Earth Shortfalls in the 2013 NDS Base Case, p. 5.

[16] According to DOD, this report was prepared in response to Section 843 of P.L. 111-83 and S.Rept. 111-201. S.Rept. 111-201 accompanied S. 3454, the Senate-proposed National Defense Authorization Act for 2011.

[17] U.S. Department of Defense. Report to Congress - Rare Earth Materials in Defense Applications. Frank Kendall, Acting Under Secretary of Defense for Acquisition, Technology and Logistics, March 12, 2012, p. 4.

[18] Ibid., p. 4.

[19] Ratnam, Gopal. Rare Earth Shortage Would Spur Pentagon to Action. *Bloomberg News*, April 9, 2012, http://www.bloomberg.com/news/2012-04-09/rare-earths-shortage-would-spur-pentagon-to-action.html.

[20] USGS Fact Sheet 087-02, and USGS Mineral Commodity Summaries, January 2010, http://pubs.usgs.gov/fs/2002/ fs087-02/.

[21] Ratnam, Gopal. Rare Earth Shortage Would Spur Pentagon to Action. *Bloomberg News*, April 9, 2012, http://www.bloomberg.com/news/2012-04-09/rare-earths-shortage-would-spur-pentagon-to-action.html.

[22] For a more detailed discussion of rare earth supply chain issues, economics, and global supply of rare earths, see CRS Report R41347, *Rare Earth Elements: The Global Supply Chain*, by Marc Humphries.

[23] U.S. Government Accountability Office (GAO). *Rare Earth Materials in the Defense Supply Chain*, GAO-10-617R, April 14, 2010, p. 19, http://www.gao.gov/new.items/d10617r.pdf.

[24] The second stage creates a rare earth concentrate that is then separated through a flotation separation process into oxide. This process is referred to as beneficiation.

[25] Cindy Hurst, "China's Ace in the Hole: Rare Earth Elements," *Joint Forces Quarterly*, Issue 59, 4th Quarter, 2010, http://www.ndu.edu/press/lib/images/jfq-59/JFQ59_121-126_Hurst.pdf.

[26] Japan produces some rare earth metal for the production of alloys and magnets for its own use.

[27] Molycorp official Andy Davis, Manager of Public Affairs, stated in May 2012 that "there has been a substantial private sector investment in rare earth projects in the United States, totaling approximately $895 million, by Molycorp alone."

[28] Confirmed on March 17, 2011, by Peter C. Dent, Vice President for Business Development, Electron Energy Corporation.

[29] Marc Humphries, CRS Specialist in Energy Policy, is the author of this section. For further discussion on recent supply chain developments, see CRS Report R41347, *Rare Earth Elements: The Global Supply Chain*, by Marc Humphries.

[30] Op. cit., Lifton Interview by The Gold Report, December 14, 2009.

[31] DOI/USGS, Rare Earth Elements—Critical Resources for High Technology, Fact Sheet, 087-02.

[32] Briefing at Mountain Pass mine site by various Molycorp Engineers, August 8, 2011.

[33] Briefing at Molycorp Metals and Alloys by Managing Director Randall Ice, on August 9, 2011.

[34] Ibid.

[35] This section of the report was written by Wayne Morrison, CRS Specialist in Trade and Finance. For the most current information on the potential impact of Chinese policies on

rare earth materials, see CRS Report R42510, *China's Rare Earth Industry and Export Regime: Economic and Trade Implications for the United States*, by Wayne M. Morrison and Rachel Y. Tang.

[36] In 2010, China produced more than 120,000 tons of rare earths, with 87,000 tons for domestic use and 34,600 tons for export 2010 (Source: the Chinese Ministry of Commerce press release, July 29, 2011.)

[37] According to the U.S. Department of Energy, China's overall export quotas declined from 65,609 in 2005 to 30,258 in 2010.

[38] *Technology Metals Research*, "How The H1-2011 Rare-Earth Export Quotas Were Allocated," December 28, 2010.

[39] *Inside U.S. Trade's World Trade Online*, U.S.," EU Denounce New China Rare Earths Export Quota As Too Restrictive", July 15, 2011.

[40] Ibid.

[41] Chinese Ministry of Commerce press release, July 15, 2011.

[42] *New York Times*, "China Said to Widen Its Embargo of Minerals," October 19, 2010.

[43] A summary of the WTO panel report can be found at http://www.wto.org/english/tratop_e/dispu_e/cases_e/ds394_e.htm#bkmk394r.

[44] The letter states that "the United States should not sit passively while China's investment policies hamstring U.S. companies and undermine our national and economic security needs."

[45] The letter can be found at http://casey.senate.gov/newsroom/press/release/print.cfm?id=81a1fa95-49d2-47a7-98b4- 65973ae14ddc.

[46] See Baron, Kevin. Pentagon Report Disputes Concerns Over China Rare Earth Reliance. *New York Times*, March 13, 2012, at http://www.nytimes.com/2012/03/09/business/global/merger-combines-a-rare-earth-mining-firm-with-aprocessor.html?_r=1&page wanted=print; Gordon, Julie. RPT-DEALTALK-Molycorp, Neo Material Deal a Rare Earth Game Changer. Reuters, March 12, 2012, at http://www.reuters.com/article/2012/03/12/neomaterial-molycorpidUSL2E8E9HTE20120312; also see Bose, Kunal. China Plays Truant With Rare Earth Metals. *Business Standard*, April 3, 2012, at http://www.business-standard.com/india/news/china-plays-truantrare-earth-metals/469896/.

[47] National Research Council. *Managing Materials for a Twenty-first Century Military, and Minerals, Critical Materials, and the U.S. Economy*, http://books.nap.edu/catalog.php?record_id=12028#toc.

[48] The Board was established through Section 843 of the John Warner National Defense Authorization Act for FY2007 (P.L. 109-364).

[49] Office of the Under Secretary of Defense for Acquisition, Technology, and Logistics. *Report of the Strategic Materials Protection Board*, December 12, 2008, http://www.acq.osd.mil/ip/docs/report_from_2nd_mtg_of_smpb_12- 2008.pdf.

[50] Ibid., p. 6.

[51] U.S. Congress. House Armed Services Committee, H.Rept. 111-166, *Report on H.R. 2647: National Defense Authorization Act for FY2010*, P.L. 111-84. The bill was signed into law on October 28, 2009.

[52] U.S. Congress. Letter to Secretary of Defense Robert M. Gates, from Senator Mark Bewitch, Senator Lisa Murkowski, and Representative Mike Coffman, January 28, 2011; http://www.politico.com/static/ PPM110_110131_rare_earth.html.

[53] U.S. Government Accountability Office. Rare Earth Materials in the Defense Supply Chain, GAO-10-617R, April 14, 2010, at http://www.gao.gov/new.items/d10617r.pdf.

[54] U.S. Department of Defense, Office of the Under Secretary of Defense for Acquisition, Technology and Logistics. *Annual Industrial Capabilities Report to Congress*, May 2010, p. 19, http://www.acq.osd.mil/ip/docs/ annual_ind_cap_rpt_to_congress_2010.pdf.

[55] U.S. Department of Energy, *Report on Critical Materials Strategy*, December 2010, pp. 58-59, http://www.energy.gov/news/documents/criticalmaterialsstrategy.pdf.

[56] U.S. Congress. House Committee on Science and Technology, Subcommittee on Investigations and Oversight, hearing on Rare Earth Minerals and 21st Century Industry, March 16, 2010, http://democrats.science.house.gov/Media/File/Commdocs/hearings/2010/Oversight/ 16mar/Hearing_Charter.pdf.

[57] NRC is the operating arm of the National Academy of Sciences, National Academy of Engineering, and the Institute of Medicine of the National Academies, chartered by Congress in 1863 to advise the government on matters of science and technology. See http://sites.nationalacademies.org/NRC/index.htm.

[58] Statement of Dr. Stephen Freiman, President, Freiman Consulting, Inc., Member, National Research Council Committee on Critical Mineral Impacts on the U.S. Economy. U.S. House of Representatives, Committee on Science and Technology, Subcommittee on Investigations and Oversight, hearing on Rare Earth Minerals and 21st Century Industry, March 26, 2010, http://democrats.science.house.gov/Media/File/Commdocs/hearings/ 2010/Oversight/16mar/ Hearing_Charter.pdf.

[59] U.S. Department of Defense. Office of the Under Secretary of Defense for Acquisition, Technology, and Logistics, Report of the Strategic Materials Protection Board, December 12, 2008.

[60] P.L. 111-383, Section 829.

[61] "USMMA Calls For A Rare Earth Strategic Reserve," *Businesswire*, February 23, 2011, http://www.businesswire.com/news/home/20110223006331/en/USMMA-Calls-Rare-Earth-Strategic-Reserve.

[62] In the first session of the 99th Congress, the role of the Critical Minerals Program was the subject of a hearing before the House Committee on Science and Technology, Subcommittee on Transportation, Aviation, and Materials. At the hearing (held October 8-10, 1985), Robert N. Broadbent, Assistant Secretary for Water and Science, U.S. Department of the Interior, testified: "The Strategic and Critical Minerals Program of the U.S. Geological Survey provides a continuing assessment of the Nation's endowment of strategic minerals and a continuing analysis of the world's mineral resources for the formulation of national minerals policy and the identification of secure sources of minerals that are critical to the security, industrial production, and economic well-being of this country and that are vulnerable to disruption in supply"; http://www.ebooksread.com/ authors-eng/aviation-united-states-congress-house-committee-on-scien/the-national-critical-materials-act-of-1984-hearings-before-the-subcommittee-o-tin/page-2-the-national-critical-materials-act-of-1984-hearings-before-the-subcommittee-o-tin.shtml. The testimony can also be viewed at http://www.archive.org/stream/nationalcritical00unit/nationalcritical 00unit_djvu.txt.

[63] Statement of Bart Gordon, Chair, House Committee on Science and Technology, Subcommittee on Investigations and Oversight, Hearing on Rare Earth Minerals and the 21st Century Industry, March 16, 2010, http://sciencedems.house.gov/publications/Opening Statement.aspx?OSID=2803.

[64] H.R. 4310, Section 901, (c), (16).

[65] H.R. 4310, Section 1604, (9), (10).

[66] P.L. 112-239 was enacted into law, January 2, 2013.

[67] Items of Special Interest, S.Rept. 112-173.

[68] Ibid.

[69] P.L. 111-83 was signed into law on January 7, 2011. Thus, the DOD report would have come due on or around July 7, 2011.

[70] Section 343, P.L. 111-83.

[71] H.R. 4402.

[72] The Task Force would be chaired by the Secretary of the Interior and would include the Secretaries (or a designee of each Secretary) of Energy, Agriculture, Defense, Commerce and State. Other members would include the Director of the Office of Management and Budget, the Chairman of the Council on Environmental Quality.

[73] H.R. 2011, Section 3. Congressional Declaration of Policy.

[74] S. 383, Section 6, Supply of Critical Minerals and Materials.

[75] P.L. 111-84 was signed into law on October 28, 2009.

[76] It should be pointed out that much of the language of the RESTART Act, proposed by Representative Mike Coffman, was included as an amendment to the FY2011 Ike Skelton National Defense Authorization Act, which was passed in the House on May 28, 2010, during the 111th Congress.

[77] The bill directs the Secretaries of Commerce, of Defense, of Energy, of the Interior, and of State to (1) appoint an Executive Agent, at the Assistant Secretary level, to serve as a representative on an interagency working group to reestablish a competitive domestic rare earth supply chain, and (2) assess and report to Congress on the chain, determining which rare earth elements are critical to national and economic security. It directs the United States Trade Representative (USTR) and the Office of Science and Technology Policy also to appoint representation to such working group. It requires the Secretary of Defense to commence procurement of critical rare earth materials and place them in a national stockpile, and the Defense Logistics Agency, Defense National Stockpile Center, to serve as administrator of the stockpile. It authorizes the administrator, if necessary to meet U.S. national security and economic needs, to purchase rare earth materials from the People's Republic of China. It instructs the USTR to (1) initiate and report to Congress on a comprehensive review of international trade practices in the rare earth materials market; or (2) initiate an action before the World Trade Organization (WTO) as a result of the review. It directs the Secretaries of Commerce, of the Interior, and of State to report to the domestic rare earth industry about mechanisms for obtaining government loan guarantees to reestablish a domestic rare earth supply chain. It directs the Secretaries of Defense and of Energy to issue guidance for the industry related to obtaining such loan guarantees. It expresses the sense of the Congress regarding a prioritization of Defense Production Act projects with respect to the domestic rare earth supply chain.

In: Rare Earth Elements
Editor: Kristin Waters
ISBN: 978-1-53610-120-1

Chapter 3

DIVERSIFICATION OF SUPPLY CHAIN AND RECLAMATION ACTIVITIES RELATED TO RARE EARTHS*

Office of the Under Secretary of Defense for Acquisition, Technology and Logistics

The Honorable Carl Levin
Chairman
Committee on Armed Services United States Senate
Washington, DC 20510

Dear Mr. Chairman:

The enclosed report responds to a request in House Report 113-102, accompanying the National Defense Authorization Act for Fiscal Year 2014. The report language requests the Under Secretary of Defense for Acquisition, Technology, and Logistics submit a report on the risk mitigation strategy used by the Department of Defense (DoD) for rare earth elements focusing on alternative sources of rare earth supply and on the reclamation of rare earths from waste.

---

* This is an edited, reformatted and augmented version of a report issued by the U.S. Department of Defense, February 2014.

The enclosed report finds that reclamation technologies for rare earth elements largely remain cost prohibitive under current conditions. DoD is actively monitoring material demand and supply, assessing risks, and considering the benefits and costs of all mitigation options to ensure a supply of rare earth materials.

Identical letters have been sent to other congressional defense committees.

Sincerely,

Frank Kendall

Enclosure: As stated

cc:

The Honorable James M. Inhofe
Ranking Member

EXECUTIVE SUMMARY

This report has been prepared pursuant to House Report 113-102, page 296, to accompany H.R. 1960, the National Defense Authorization Act for Fiscal Year 2014. The House Report requires DoD to submit this report on the Department's earlier established risk mitigation strategy[1] for rare earths for U.S. defense demands, focusing on alternative sources of rare earth supply and in particular the reclamation of rare earths from waste.[2]

Overall, the Department's strategy in this area consists of four parts: (1) active monitoring of material demand and supply conditions; (2) structured risk assessments; (3) careful consideration of the benefits and costs of various risk mitigation options; and (4) a three-part risk mitigation strategy that includes diversification of supply, pursuit of substitutes, and reclamation of waste.[3] The constant, active monitoring of the global situation concerning rare earths is the first and very important part of this overall strategy, as is the regular assessment of risks to the DoD supply of rare earths under various planning cases and conditions. The last major parts of the overall strategy are structured consideration of actions to be taken to mitigate risks as the situation warrants, and selective DoD risk mitigating initiatives, as appropriate. Today, DoD is implementing all major parts of its strategy and remains ready to take further actions using particular mitigation options as needed.

The Department observes that market supply and demand conditions overall have significantly improved in the United States and internationally since the 2010 and 2011 turmoil in the rare earths markets.[4] As cited in this report, there has been a significant decrease in the demand and prices for rare earths, as well as an increase in the availability of supply, including material surpluses and diversification in production (e.g., re-opening of major rare earths mining operations and magnet production in the United States).

In addition to an 11.8-percent decline in the overall global demand for rare earths from 2010 to 2013.[5] earlier forecasted growth in demand through 2016 has been revised downward.[6] The Department of Defense projects almost no growth in U.S. defense demand for rare earths between 2015 and 2018.[7]

The Department observes that the supply of rare earths for U.S. defense acquisition programs is not presently being disrupted. The Department also estimates no shortfalls of rare earth supplies of ores and concentrates for U.S. defense demands through the 2015 to 2018 timeframe.[8]

With respect to rare earth reclamation, DoD evaluated the applications for which it uses rare earths and the potential for the recovery of rare earths from those applications at end-of-product-life (e.g., magnets, batteries, displays and light bulbs). The Department also evaluated the current potential for reclaiming rare earths from tailings of U.S. mining operations. However, many reclamation approaches are largely considered cost-prohibitive in the United States because of a variety of factors including, among others, technical challenges, economic barriers, and regulatory issues. Within the private sector, with the exception of the partial recovery of rare earths from phosphors contained in light bulbs, there are no commercial-scale rare earth recycling and reclamation operations in the United States, and limited operations of this kind exist outside the country.

Chiefly for these reasons, DoD does not currently see a compelling need to take programmatic action with respect to rare earth recycling and reclamation and indeed, because of the lack of a sufficient business case, the Department does not currently have any operational-level rare earth recycling and reclamation defense programs. Nevertheless, DoD continues to support selected promising rare earth materials research and assessment activities, including reclamation and recycling. Other U.S. Government agencies (e.g., the Department of Energy and the National Science Foundation) are doing the same, consistent with their missions.

Moving forward, and as a regular part of its larger strategy in this area, DoD continues to actively monitor rare earth markets, assess potential risks to its supply, and evaluate U.S. defense needs, risk mitigation options, and related DoD program requirements. Along these lines, in the last 3 years DoD conducted and reported to Congress on six separate assessments of various aspects of rare earth supply risk.[9] Those assessment efforts led the Department to undertake further risk mitigation measures. DoD remains prepared to take additional actions on its rare earth risk mitigation strategy as circumstances warrant. In the meantime, the U.S. defense industrial base is prepared to take advantage of future rare earth market opportunities as they arise (e.g., increased supply opportunities that may result from private sector progress in rare earths reclamation).

INTRODUCTION

This report has been prepared pursuant to House Report 113-102 that accompanied H.R. 1960, National Defense Authorization Act (NDAA) for Fiscal Year 2014 (See Appendix 1). The House Report requests the Under Secretary of Defense for Acquisition, Technology, and Logistics (USD(AT&L)) to submit a report to the congressional defense committees by February 1, 2014, on the Department of Defense's risk mitigation strategy for rare earth elements as it pertains to U.S. defense demands. The House Report states its understanding that the Department intends to pursue a three-pronged strategy to secure supplies of rare earth elements, which consists of diversification of supply, pursuit of substitutes, and a focus on reclamation of waste, as part of a larger U.S. Government recycling effort. The House Report requests the Under Secretary's Report to answer several questions focused on alternative sources of rare earth supply from the reclamation of waste.

DoD Risk Mitigation Strategy for Rare Earth Elements

Due to a combination of reduced rare earth supplies from China coupled with a temporary sharp rise in demand for related materials in 2010-2011, the rare earth market was exhibiting large price spikes and showing some potential for shortages to occur within the United States, with respect to both raw materials and products containing rare earths. In the Ike Skelton National Defense Authorization Act for Fiscal Year 2011, Public Law 111-383, Congress required the Department to prepare a report on the supply and demand for rare earth materials in U.S. defense applications. In that report, DoD stated its risk mitigation strategy for rare earths.[10] That strategy, which DoD is currently pursuing, includes the aforementioned three-prongs: diversification of supply, pursuit of substitutes, and a focus on reclamation of waste as part of a larger U.S. Government recycling effort. Diversification means taking advantage of new sources of rare earths, including virgin material from new mines, materials and products from new factories, material saved via more efficient processing and manufacturing technologies (e.g., saving of rare earth containing manufacturing scrap), and material conserved via new product designs. Substitution means taking advantage of direct, material-for-material substitutes as well as functional substitutes—products that perform the same functions as products using rare earths but via different technical approaches. Reclamation of waste includes using rare earths recovered from mine tailings, recycled components post-consumer waste, and other waste streams such as those from industrial production of other materials.

The first part of the Department's strategy in this area consists of the constant, active monitoring of the global situation concerning rare earths, the assessment of its implications for the supply of rare earths to the defense industrial base, and the consideration of actions to be taken to mitigate risks as the situation warrants. DoD monitors the rare earth markets, to include the global supplies of and demands for materials and products. It assesses potential risks to the defense and essential civilian supply chains necessary to assure secure supplies of rare earths essential to national security. It also considers actions to take that include developing policy, contingency plans, and other instruments to ensure the availability of supply. These monitoring, assessing, and planning activities are carried out regularly by various offices within the Office of the Secretary of Defense and the Military Departments.

Although rare earth materials are widely used within the defense industrial base, such end uses represent a small fraction of total U.S. consumption. Therefore, as it does with most materials, DoD looks first to the market to mitigate rare earth supply risks. Indeed, the market has been responding to mitigate the risks that were seen at the time DoD stated its rare earth risk mitigation strategy (e.g., developing new and diverse sources of supply and reducing demand by developing substitutes and product designs that use less or no rare earth material). However, where conditions warrant, DoD takes direct action to mitigate such risks further. Such actions can include investing in assured production or materials processing capabilities, establishing secure inventories of materials or parts, and/or establishing additional qualified defense material suppliers.

There have been limited economic incentives within the marketplace to reclaim and recycle rare earths. Furthermore, because U.S. defense demand for rare earths is small relative to their estimated supply, DoD lacks the necessary requirements that would warrant programmatic actions (current or planned) to execute those aspects of its rare earth strategy for implementing commercial-scale reclamation of rare earths—including recycling of end-of-life products (e.g., electronic devices) and/or processing of rare earth containing tailings from earlier mining of minerals.

The basis for this decision is reflected in DoD's Strategic and Critical Materials 2013 Report on Stockpile Requirements that was coordinated by the Department and submitted to the Congress in January 2013. In that report the DoD evaluated Department-wide U.S. defense and essential civilian demands for rare earths (along with numerous other materials) for 2015 through 2018. For this purpose it assessed the availability of global rare earths supply in the context of a Congressionally-mandated, and large scale, national emergency that included major conflict scenarios and significant rare earths supply disruptions. Of the rare earth ores and concentrates assessed for the 2013 Report on Stockpile Requirements, there were no projected shortfalls estimated for U.S. defense purposes—including rare earths requirements for defense acquisition programs and the associated indirect demands for rare earths necessary to support the U.S. defense industrials base.[11] There was a special National Defense Stockpile (NDS) supply chain assessment for one "downstream" rare earth material that identified a shortfall for a very small amount of a high purity "heavy" rare earth oxide.[12]

In terms of peacetime U.S. defense demands for rare earths, and at the individual defense program level (e.g., individual weapon system), during the past few years there have been no major supply interruptions that have significantly disrupted DoD defense programs. Although DoD had noted reports of temporary price increases and possible tight inventories of some rare earth materials at the height of the rare earth market turmoil from late 2010 through 2011, there were no major supply disruptions to DoD's defense acquisition programs during this period.

Based on DoD's 2013 Report on Stockpile Requirements to the Congress and the fact that there is no evidence of major interruptions to the supply of rare earths that have disrupted current or recent defense acquisition programs, there are no U.S. defense requirements—currently or estimated over the next 5 years—that would warrant DoD initiating commercial-scale rare earth reclamation defense programs as a means to mitigate existing or anticipated rare earths shortfalls.

A decision by DoD to directly execute any of its rare earth supply risk mitigation strategies (i.e., traditional government stockpiling in addition to establishing U.S. defense specific programs in reclamation, substitution, and diversification of supply) would be driven by market (and geopolitical) supply and demand considerations—specifically, situations where U.S. defense industrial base requirements for rare earths could not be satisfied by the private sector as is done under current U.S. defense acquisition practices and industrial base conditions. Conditions that could lead to DoD undertaking extraordinary steps to intervene directly in private sector rare earth markets in support of U.S. defense needs include potential rare earth supply disruptions that would be severe and persistent enough to seriously affect DoD's defense acquisition programs and threaten serious harm to U.S. military capabilities.

To prepare for the unlikely event of a major interruption of the supply of rare earths for U.S. defense requirements (or other significant market developments that could seriously affect U.S. defense acquisition programs and military capabilities), DoD's contingency planning activities include comprehensive assessments of supply and demand, as well as ongoing evaluation of potential supply risks and mitigation options.

RECENT RARE EARTH MARKET DEVELOPMENTS

Since DoD's initial rare earths report to the Congress in 2011,[13] market conditions—reflected by an increase in supply and a decrease in demand—have significantly improved[14] for consumers of rare earths, including suppliers within the U.S. defense industrial base. Improvements in the supply of rare earths that have benefited consumers include the largest and most commonly used rare earths: light and medium rare earth elements including those used to produce catalysts, polishing powders, glass, batteries, and magnets. There have also been relative improvements in the availability of much less commonly used (albeit still very important) heavy rare earth elements including those used mainly in the production of phosphors for lighting products and displays, as well as those used in smaller quantities as additives within a variety of specialized materials including crystals for lasers, dopants in magnet alloys, and stabilizers for high temperature jet engine component coatings and ceramics for fuel cells.

Examples of significant market changes that have benefited consumers of rare earth materials since late 2010 and mid 2011 include:

- Prices for basic rare earth materials (e.g., compounds) have fallen significantly since their historic highs of the last 2 to 3 years—including the most widely used light and medium rare earth elements as well as less used heavy rare earths.
- The available supply of rare earths has significantly improved for the most commonly used rare earth elements (including those used to produce catalysts, batteries, polishing material, glass and magnets). In some cases, the most widely used rare earths are deemed by producers to be in surplus.[15] The available supply of even less commonly used rare earths—specifically heavy rare earths—has also improved during the past few years.
- There have been significant supply side improvements in global rare earth markets that have further benefited consumers, including a number of very positive private sector developments in North America. Most significant has been U.S.-based Molycorp's recently re-opened Mt. Pass rare earth mine and processing plant in California (with a total rare earth element oxide annual production capacity of 19,050 metric tons). The company has continued to expand and diversify its other rare earth production capabilities in

the United States and internationally. Actions include acquisitions, joint ventures, and other key supplier arrangements throughout rare earth supply chains in Asia, Europe, and North America.

- Another very significant supply side development for U.S. buyers and consumers of rare earths involves a recent rare earth magnet production plant opening in the United States, a plant owned by the Japanese-based firm Hitachi. This new manufacturing facility in North Carolina has production capacity and capabilities to produce large quantities (relative to the needs of the defense industrial base) and diverse grades of high-performance, sintered Neodymium Iron Boron (NdFeB) rare earth magnets. In addition to this new U.S. plant's ability to support important U.S. commercial and industrial market demands for sintered NdFeB magnets, the facility could serve as a key supplier within DoD's U.S. defense industrial base.
- Other positive rare earths supply side developments underway in North America include the ongoing exploration and development of several new rare earth mining and processing ventures in the United States and Canada, including potential possibilities to produce rare earths in the future from tailings generated from earlier mining operations.
- In addition to an 11.8 percent decline in the overall global demand for rare earths from 2010 to 2013,[16] earlier forecasted growth in demand through 2016 has been revised downward." The Department projects almost no growth in U.S. defense demand for rare earths between 2015 and 2018.[18]
- The demand for rare earths has also changed in a number of significant ways that has provided additional benefits to buyers and consumers of these materials. Specifically, important segments of the demand for rare earths have essentially been eliminated or greatly diminished, easing the overall demand for rare earths. Pressures on the demand for rare earths have declined because of a number of factors, including private sector market responses to develop substitutes for rare earths as well as other efforts to reduce the demand for rare earths. Examples include designing products that are less dependent on rare earths (e.g., reduced use of heavy rare earth dopants in magnets), replacing the use of scarce heavy rare earths with more widely available light rare earths, re-using rare earth polishing materials, and, in certain cases, doing away with the use of rare earths in products all together (e.g., increased use of LED

lighting, Li-ion batteries, and ferrite magnets that use much less or no rare earth elements).

- Different industry sectors that make significant use of rare earths in manufacturing have found ways to either increase internal production efficiencies so as to ultimately reduce rare earth containing scrap (e.g., generated from the production of new magnets) or reintroduce into their production processes rare earths recovered from recycling end-of-life products (e.g., rare earth recovered from phosphors used in discarded light bulbs).
- Additionally, there are a variety of U.S. as well as international industry, academic, and government-sponsored rare earths materials and processing research activities underway. Regarding the latter, U.S. government agencies, including DoD, are funding rare earths research projects in rare earths reclamation such as recycling rare earths from end-of-life consumer products (magnets) and industrial waste (fly ash) as well as processing rare earth containing tailings from earlier mining operations.

RARE EARTH RECYCLING AND RECLAMATION

Federal Programs

Because of favorable rare earth market conditions for consumers of rare earths, the lack of sufficient U.S. defense requirements (as reported in DoD's recent 2013 Report on Stockpile Requirements and as evidenced by a lack of rare earth supply interruptions that have significantly disrupted major U.S. defense programs and military capabilities), and the relative lack of economic viability for rare earth reclamation and recycling in the United States, DoD does not have any operational-level (i.e., commercial-scale) rare earth reclamation programs (current or planned).

Nevertheless, DoD is actively assessing current and future rare earth supplies and Department-wide demands (currently through 2020 for the next Report on Stockpile Requirements to the Congress), and DoD has made investment resources available for rare earth materials and processing research and development (R&D). Research has focused on supporting advancements in improved rare earths materials processing and recycling. Examples of DoD research funding have included industry research funding for rare earth metal refining by the Army and early-stage small business R&D funding by the Navy for reclaiming rare earths from waste material streams. Other DoD R&D programs have been made available to support rare earth materials process development: Defense Logistics Agency (DLA) Strategic Materials' research Broad Agency Announcement, the Title III program of the Defense Production Act, the Defense Manufacturing Technology program, and DoD's Rapid Innovation Program.

DoD will continue to support these and related rare earth materials and processing R&D activities—including those specifically related to rare earths reclamation technologies—as defense research opportunities and related U.S. military requirements dictate.

Beyond the DoD R&D activities noted above, other federal agencies, primarily the Department of Energy (DoE) at its Critical Materials Institute and the National Science Foundation, are supporting research programs into rare earth recycling and reclamation (see https://cmi.amerslab. goviresearch/improving-reuse-recycling). The Critical Materials Institute was established by DoE to help the United States avoid the impact of supply availability concerns for certain materials, including rare earths, by developing technologies that will make better use of the materials to which we have access as well as eliminate the need for materials that are subject to supply disruptions. It is sponsoring several research projects related to rare earth reclamation, including: the recovery of rare earth metal oxides from lamp phosphor dusts; the electro-chemical recycling of rare earth metals from electronic wastes; and the study of the logistics and economics of rare earth remanufacturing and recovery. The National Science Foundation is sponsoring several projects at colleges and universities related to rare earth reclamation, including the study of: biological and chemical technologies related to rare earths and critical materials; advanced manufacturing and nanotechnology for rare earths and critical materials; and magnet separation technologies for recycling. The Environmental Protection Agency also is sponsoring research concerning the cost-effective recycling of rare earth magnetic alloys and permanent magnets from industrial scrap and discarded electronic products.

These government programs are developmental because, with the exception of the two examples of the reclamation of material from fluorescent light phosphors and NiMH batteries outside of the United States, the current state of the technology for rare earth recovery does not allow material to be reclaimed economically (in the North American market) from products at the ends of their lives. While the technology exists to recover rare earths from several other sources (e.g., NdFeB magnets in computers and electronic components, thermal barrier coatings in turbine engines, and automotive catalytic converters), such recovery does not appear economical under current market conditions for these application areas.

The situation is somewhat different in Europe, where more stringent environmental laws require the collection and processing of certain rare earth-bearing waste products and thereby make the recovery of rare earths more attractive there. Two such product types from which rare earths are recovered on a commercial scale today are fluorescent lights in France and NiMH batteries in Belgium.[19] In Japan, rare earths are being recovered, on a limited scale and potentially developmental basis, from magnets from computers, consumer electronics, and household appliances, and from NiMH batteries from hybrid cars.

While DoD is not choosing to take commercial scale programmatic actions on the reclamation prong of its rare earth risk mitigation strategy right now (i.e., beyond limited research activities), DoD is prepared to take advantage of existing private sector capabilities as well as any future capabilities that may result from industry and other R&D programs that could help meet U.S. defense demands for rare earths as circumstances change and DoD's requirements dictate (e.g., anticipation of significant interruptions of future supply and that would result in major disruptions to U.S. defense programs and military capabilities).

As noted above, the DoD and other U.S. government programs concerning rare earth recycling and reclamation are largely in the research and development stage. Their goal is to develop cost-effective recycling and reclamation capabilities. Researchers estimate that they will lead to commercialization mostly in the mid to longer-term, roughly 5 to 7 years from now. Therefore, this report cannot state the costs of the rare earth materials that they will produce compared to the cost of mined or newly produced materials.

Reclamation Resources Investigated

DoD has investigated potential alternative sources of supply of rare earths in several respects. In 2012, the Secretary of Defense submitted a report to Congress on the feasibility and desirability of recycling, recovering, and reprocessing rare earth elements.[20] The study that produced the report investigated the following major DoD applications for rare earths for their potential for recycling and recovery:

- Neodymium-iron-boron (NdFeB) magnets
- Samarium-cobalt (SmCo) magnets
- Phosphors used in displays
- Phosphors used in lighting
- Ceramics and glass
- Fiber optics
- Nickel-metal hydride (NiMH) battery alloys
- Metal alloys
- Lasers

Those applications were evaluated based on the availability of the rare earth materials in various systems and forms; the state of the materials processing technologies and collection, storage, and transportation infrastructures; and the economics of the materials recovery processes. The following five application areas were identified as offering the greatest theoretical potential for recycling and recovery: phosphors for lighting and visual displays, NdFeB magnets in computers and electronic components, NiMH battery alloys, high-temperature ceramics and ceramic coatings from turbine engines, and automotive catalytic converters.

In addition to that effort, DoD (the Office of Naval Research) is currently sponsoring research into the potential reclaiming rare earth elements from coal fly ash. It (DLA Strategic Materials and the Navy) is also exploring the feasibility of recycling rare earth laser crystals at the end of the life cycle.

In evaluating DoD rare earth application areas for their potential for recycling and recovery, the 2012 DoD study estimated the quantity and form of rare earth elements that could theoretically be recovered by recycling products after the ends of their lives. The estimates were theoretical in that in some cases the technology of recycling and recovery remains challenging, and for all cases (with the exceptions of lamp phosphors and NiMH batteries noted above), the market has not yet found rare earth recycling and recovery to be economical, in that there are no commercial scale rare earth recycling and recovery operations in North America today. The estimates of potential material availability from DoD applications were as follows.

Table 1. 2012 Estimate of Rare Earths Theoretically Recoverable from DoD Applications

Product Type	Rare Earth	Potential Annual Amount (MT)
Fluorescent Lamps	Yttrium oxide	58
	Europium oxide	4
	Erbium oxide	3
Nickel Metal Hydride Batteries	Lanthanum	3.5 to 5
	Neodymium	3.5 to 5
Neodymium Magnets from	Neodymium	3.5
Computers and Other Electronic Components	Praseodymium	1
	Dysprosium	0.25
	Gadolinium	0.1
	Terbium	0.01
Thermal Barrier Coatings	Yttrium oxide	5
Automotive Catalytic Converters	Cerium oxide	20

Mine Tailings

In addition to rare earth-containing wastes that could potentially be reclaimed (pending technological and/or market developments), there are also mine tailings in the United States that contain rare earths that could potentially serve as a source of additional supply.[21] At the Mountain Pass mine in California, the tailings basin contains up to 5 percent rare earth oxide (in monazite) and the tailings mounds contain approximately 2 to 5 percent rare earth oxide. The total amount of rare earths contained in the Mountain Pass tailings is significant—they could be characterized as the second largest deposit of rare earths in North America (next to the deposits in the ground at Mountain Pass).[22] Tailings associated with the Pea Ridge, Missouri, iron ore body and mine and the former iron mines near Mineville, New York, also contain significant quantities of rare earths. There are also tailings associated with some other mineral developments (e.g., phosphate minerals) that contain rare earths. The U.S. Geological Survey is continuing to evaluate potential domestic sources of rare earths. DoE is sponsoring some research into the development of technologies that could make the recovery of materials from mine tailings more effective. Future exploitation of tailings resources will depend on technological, regulatory, and/or market developments.

The Use of the Defense Priorities and Allocations System for Rare Earth Materials

In addition to questions concerning rare earth recycling and reclamation, the House Report also requests a description of the materials that the Department plans to obtain via the Defense Priorities and Allocations System (DPAS). The DPAS is typically applied at a program-level basis rather than to individual raw materials. Once a program is given a DPAS rating, the rating is extended to all U.S. contracts and subcontracts within the program, including material suppliers if applicable. The Unite States has no ability to rate a contract from a foreign supplier unless a Security of Supply Arrangement has been established with the foreign country and the supplier has agreed to comply with the Arrangement. Most DoD programs have a DPAS rating.[23] Exceptions are some non-critical items like commonly available commercial goods and goods that do not directly support logistics, tactical, or operational program requirements.[24] It is estimated that 300,000 contracts and purchase orders annually include priority ratings. The Department does not track which of those orders maybe related specifically to rare earth materials.

Potential Conflict Risk and Mobilization and Other Risk Mitigation Measures

DoD assesses rare earths supply risks in the context of both current peacetime conditions and mobilization and potential large scale conflict scenarios. As noted above, DoD has not identified major risks of severe disruptions under current peacetime conditions of the rare earths supply for the U.S. defense industrial base generally or for specific DoD defense acquisition programs. DoD's Department-wide assessment of rare earths supply risks within the context of a major national emergency (including large scale conflicts) is conducted by DoD's National Defense Stockpile (NDS) program and the NDS biennial Stockpile Requirements Report process. The NDS program and its requirements development process is DoD's primary Department-wide means for identifying and mitigating future risks to the anticipated supply of strategic and critical materials—primarily at the raw materials level of ores and concentrates and including rare earth elements.[25] DoD's requirements are determined for all U.S. defense demands and essential civilian demands within the context of a congressionally-mandated national emergency scenario.

With the exception of a small quantity of one specialty rare earth compound, DoD's most recent Report on Stockpile Requirements to the Congress estimated no shortfall of rare earths materials for U.S. defense needs during a postulated national emergency scenario in the 2015 to 2018 time period.[26]

That said, DoD recognizes potentially unique risks to the supply of selective rare earth materials, such as certain heavy rare earth elements; related separation and purification production capabilities; and the capabilities and capacities to produce downstream value added materials and associated products including various rare earth compounds, metals and alloys.[27] For these purposes, DoD has been undertaking extensive "downstream" supply chain assessments of various rare earth containing materials (e.g., magnets, ceramics, structural alloys, specialty metals and chemicals) that may be of particular importance to U.S. defense acquisition programs and the military capabilities these programs support.

For these and other reasons, DoD is in the process of establishing an initial stockpile of select rare earth containing materials (including refined metals and specialty chemicals). DoD is additionally pursuing an expansion of the legal authorities for the NDS program to further utilize recycled materials for stockpiling and related purposes, which will help enable DoD to take greater advantage of future market developments in rare earth reclamation.

As a final matter regarding the mitigation of risk associated with rare earths, it should be emphasized that in addition to the discrete actions discussed above, DoD's strategy includes broadly the constant, active monitoring of the global situation concerning rare earths, the assessment of its implications for the Department's supply of rare earths, and the consideration of actions to be taken to mitigate risks as the situation warrants. DoD has been paying increasing attention to rare earths as market developments have suggested the existence of possible threats to its supply:

- In 2011, OUSD(AT&L) presented its 2011 Report To The Congress On National Defense Stockpile Requirements in which it evaluated the supply of and demand for seven rare earth elements under its statutory NDS planning scenario (and others) and recommended them for inclusion in the Stockpile (to mitigate shortfalls of material needed for essential civilian demands (no defense shortfalls were identified)).

It also discussed alternatives to stockpiling for mitigating material shortfalls.

- In January 2012, OUSD(AT&L) presented its aforementioned report, Rare Earth Materials in Defense Applications, in which it assessed the supply of and the defense applications for all 17 rare earth elements and stated its rare earth risk mitigation strategy. That report noted the forthcoming construction of the Hitachi neodymium-iron-boron magnet plant in North Carolina (which is now operating) as a market response mitigating a vulnerability in an important U.S. rare earth supply chain. This document estimated that only one rare earth at the raw material level, yttrium, would be in shortfall for U.S. defense purposes.
- In September 2012, OUSD(AT&L) presented its aforementioned Report on Feasibility and Desirability of Recycling, Recovery, and Reprocessing Rare Earth Elements, in which it evaluated the potential for rare earth recycling to mitigate supply risks.
- In December 2012, the Office of the Secretary of Defense presented its Report to Congress on Assessment of Feasibility and Advisability of Establishment of Rare Earth Material Inventory, in which it assessed the feasibility and advisability of establishing an inventory of rare earths to assure their long-term availability (recommending ultra-pure yttrium oxide for addition to the Stockpile) and stated the further actions the Department intended to take to assure its rare earth supply. That report did not identify further shortfalls for U.S. defense requirements at the raw material level. However, uncertainties regarding the supplies of certain heavy rare earth raw materials, higher purity oxides, related compounds, and semi-finished products made it advisable to maintain a high level of surveillance, assess and establish mitigation solutions of semi-processed rare earth containing materials, and maintain larger inventories of spares for critical systems.
- In January 2013, OUSD(AT&L) presented its Strategic and Critical Materials 2013 Report on Stockpile Requirements, in which it assessed risks arising from the supply of and demand for all 17 rare earth elements under the NDS planning scenario and recommended four for stockpiling to cover essential civilian shortfalls. It also discussed in more depth alternatives to stockpiling for mitigating material shortfall risk. With the exception of a relatively small quantity of a specialized rare earth compound, this report did not

estimate any rare earth material shortfalls for U.S. defense specific needs.

- In October 2013, OUSD(AT&L) presented its most recent Annual Industrial Capabilities Report to Congress, in which it discussed recent developments in the rare earths markets in the United States and overseas, including efforts made by the market and by the Department to reduce rare earths supply risks.
- DLA Strategic Materials is now assessing, for the FY 2015 Report on Stockpile Requirements, the risks arising from the supply of and demand for the rare earths under the NDS planning scenario. It is extending its material assessments to include risks arising from vulnerabilities in selected global supply chains (downstream of raw material production) for defense and essential civilian demands. It is also continuing to develop its approach to evaluating alternatives to stockpiling for mitigating material shortfall risk.

Thus, DoD has made considerable efforts since 2011 to assess risks to the Nation's supply of rare earths and will continue to do so into the future as part of pursuing its risk mitigation strategy.

CONCLUSION

In response to the House Report 113-102 request for this report on DoD's risk mitigation strategy for rare earth elements and the House Report's focus on the reclamation of waste, the Department concludes as follows:

- It is the Department's overall strategy to ensure the supply of rare earth materials by: (1) actively monitoring material demand and supply; (2) assessing risks; and (3) considering the benefits and costs of mitigation options.
- In addition to constant, active monitoring of the global situation concerning rare earths, the Department's specific three-part strategy for mitigating the potential risk to the supply of rare earths includes: (1) leveraging diversification of supply, (2) taking advantage of substitutes, and (3) using materials available from the reclamation of waste. DoD is implementing this overall strategy for rare earth

materials, and remains ready to take further risk mitigation actions using particular options as appropriate.

- The Department observes that the peacetime supply of rare earths for U.S. defense acquisition programs is not presently being disrupted and it expects no shortfalls of rare earths for U.S. defense demands through the 2015 to 2018 timeframe—even in the context of a major national emergency when some supply interruptions would be expected to occur.
- From the perspective of users of rare earths, market supply and demand conditions overall have significantly improved in the United States and internationally since 2010 and 2011. There has been a significant decrease in the demand and prices for rare earths as well as an increase in the availability of supply including material surpluses and diversification in production.
- DoD recognizes potential risks involving certain aspects of rare earth supply, such as some heavy rare earths, downstream separation and purification capabilities, and the capabilities and capacities to produce further value added materials and associated products. To address these risks, DoD has been undertaking extensive "downstream" supply chain assessments of rare earth containing materials that may be of particular importance to U.S. defense programs, and where appropriate, initiated risk mitigation strategies (e.g., received recent authorization by the Congress to stockpile a specialty rare earth oxide for lasers used in U.S. military systems and investigating re-use of related materials from old military systems).
- With respect to rare earth reclamation, DoD has evaluated the applications for which it uses rare earths and the potential for the recovery of rare earths from those applications. DoD observes that reclamation technologies largely remain immature and/or otherwise cost prohibitive, and thus, there is limited to no commercial-scale recycling of post-consumer waste or mine tailings in North America.
- For the foregoing reasons, DoD has not yet seen a compelling need to take programmatic action with respect to commercial scale rare earth recycling and reclamation, and it does not currently have any operational-level recycling and reclamation defense programs. DoD is supporting, and will continue to support, rare earth materials research and related assessments, including those concerning recycling and reclamation. Where appropriate, it will leverage related research activities supported by other U.S. agencies, industry, and academia.

- Moving forward, and as a regular part of its larger strategy in the area of strategic and critical materials risk mitigation, DoD will continue to actively monitor rare earth markets, assess potential risks to its supply, evaluate U.S. defense needs, consider risk mitigation options, and related DoD program requirements. Those assessment efforts have led the Department to undertake further risk mitigation measures.
- DoD remains prepared to take additional actions on its rare earth risk mitigation strategy as circumstances warrant while the U.S. defense industrial base is prepared to take advantage of future rare earth market opportunities as they may develop (e.g., increased supply opportunities that may later develop from private sector progress in rare earths reclamation).

APPENDIX 1: FISCAL YEAR 2014 NATIONAL DEFENSE AUTHORIZATION ACT, HOUSE REPORT 113-102

National Defense Authorization Act for Fiscal Year 2014 Report of the Committee on Armed
Services House of Representatives
TO ACCOMPANY
H.R. 1960
June 7, 2013

Report on Diversification of Supply Activities Related to Rare Earth Elements

The committee is aware that in response to the report required by section 843 of the Ike Skelton National Defense Authorization Act for Fiscal Year 2011 (Public Law 111-383) and based on forecasting demand for fiscal year 2013 only, the Under Secretary of Defense for Acquisition, Technology, and Logistics concluded that domestic production of rare earth elements could satisfy the level of consumption required to meet defense procurement needs by fiscal year 2013, with the exception of yttrium. However, the committee observes that the Future Years Defense Program indicates that consumption of rare earth elements is expected to increase after 2013. Specifically, the report on the feasibility and desirability of recycling, recovery, and reprocessing of rare earth elements required by the conference report (H. Rept. 112-329) to accompany the National Defense Authorization Act for Fiscal Year 2012, states that each SSN-774 Virginia-class submarine would require approximately 9,200 pounds of rare earth materials, each DDG51 Aegis destroyer would require approximately 5,200 pounds of these materials, and each F-35 Lightning II aircraft would require approximately 920 pounds of these materials.

The committee is aware that the Department of Defense intends to pursue a three-pronged strategy to secure supplies of rare earth elements, which consists of diversification of supply, pursuit of substitutes, and a focus on reclamation of waste, as part of a larger U.S. Government recycling effort. The committee believes that diversification of supply activities related to rare earth elements is necessary in order to meet the growing demand for these materials, but the committee is concerned that some of these processes may prove to be technically difficult or so expensive that they are deemed cost-prohibitive.

Therefore, the committee directs the Under Secretary of Defense for Acquisition, Technology, and Logistics to submit a report to the congressional defense committees by February 1, 2014, on the Department's risk mitigation strategy for rare earth elements, which should include, at a minimum, the following elements:

(1) A list and description of the programs initiated or planned to reclaim rare earth elements by the Department, along with a description of the materials reclaimed or expected to be reclaimed from such programs;
(2) An assessment of the cost of materials produced by these reclamation efforts compared to the cost of newly-mined materials;
(3) An assessment of availability of reliable suppliers in the National Defense Industrial Base for the reclamation and reprocessing of rare earth elements;
(4) A list of alternative sources of supply, such as mine tailings, recycled components, and consumer waste, that the Department has investigated or plans to investigate;
(5) A physical description of alternative sources of supply with corresponding geologic characteristics, such as grade, resource size, and the amenability of that feedstock to metallurgical processing;
(6) A description of the materials that the Department plans to obtain via the Defense Priorities and Allocations System;

(7) Other diversification of supply activities deemed relevant by the Under Secretary.

End Notes

[1] This strategy was developed by the Department of Defense pursuant to Section 843 of the National Defense Authorization Act for FY 2011 (P.L. 111-383) and Senate Report 111-201.

[2] For the purposes of this report, reclamation includes the recovery of material from end-of-life manufactured products that contain rare earth elements (e.g., magnets from old computer hard drives, phosphors from discarded light bulbs and batteries from consumer electronics) as well as recovering rare earths from tailings produced from earlier mining activities. While this report mentions opportunities that exist to reprocess rare earth containing scrap generated from producing new products (e.g., rare earth magnets) and the increased reuse of rare earth containing industrial manufacturing materials (e.g., rare earth based polishes), both areas of rare earths conservation were considered outside of the primary focus of House Report 113-102 on recovering rare earths from discarded waste materials and products.

[3] The Department also has additional risk mitigation authorities and mechanisms available (e.g., Strategic and Critical Materials Stock Piling Act and the Defense Production Act).

[4] Annual Industrial Base Capabilities Report to Congress, submitted by the Under Secretary of Defense for Acquisition, Technology and Logistics, Office of the Deputy Assistant Secretary of Defense for Manufacturing and Industrial Base Policy, October 2013.

[5] Curtin-IMCOA, Rare Earths Quarterly Bulletin #5, by Dudley J. Kingsnorth, June 2013.

[6] The Global Rare Earths Industry: Poised for Growth, by Dudley J. Kingsnorth, November 2012.

[7] Strategic and Critical Materials 2013 Report on Stockpile Requirements.

[8] Supply shortfall assessments of U.S. defense demands for rare earth ores and concentrates for the period of 2015 through 2018 are reported in the Strategic and Critical Materials 2013 Report on Stockpile Requirements submitted to the Congress by the Department of Defense in January 2013. This report did estimate a U.S. defense shortfall for a specialty rare earth oxide used in lasers for certain weapon systems. To mitigate this estimated shortfall, DoD has recently been authorized by the Congress, in the FY 14 NDAA, to create a stockpile of related materials. In addition, DoD is actively investigating the potential for re-using rare earth containing lasers from end-of-life of military equipment.

[9] Rare Earth Elements in National Defense: Background, Oversight Issues, and Options for Congress, by Valerie Bailey Grasso, September 17, 2013.

[10] Under Secretary of Defense for Acquisition, Technology and Logistics, Report to Congress, Rare Earth. Materials in Defense Applications (March 2012); see also Office of the Secretary of Defense, Report to Congress on Assessment of Feasibility and Advisability of Establishment of Rare Earth Material Inventory (September 2012).

[11] The 2013 Report on Stockpile Requirements identified rare earth supply shortfalls for essential civilian demands during the report's 2015 - 2018 Base Case scenario: yttrium (1,899 MTs), scandium (572 MTs), erbium (124 MTs), dysprosium (47 MTs), thulium (20 MTs) and terbium (7 MTs).

[12] The 2013 Report on Stockpile Requirements also reported U.S. defense shortfalls for beryllium and several high performance carbon fibers.

[13] Interim Report, Assessment and Plan for Critical Rare Earth Materials in Defense Applications, from USD (AT&L) to the Congress, August 2011.

[14] The Global Rare Earths Industry: Poised for Growth, by Dudley J. Kingsnorth, November 2012.

[15] The Global Rare Earths Industry: Poised for Growth, by Dudley J. Kingsnorth, November 2012.

[16] .Curtm-IMCOA, Rare Earths Quarterly Bulletin #5, by Dudley J. Kingsnorth, June 2013.

[17] The Global Rare Earths Industry: Poised for Growth, by Dudley J. Kingsnorth, November 2012.

[18] 2013 Report on Stockpile Requirements.

[19] The European firm that reclaims rare earth oxides from NiMH batteries also collects batteries in the United States for recovery of the material in Europe. The European firm that reclaims rare earths from fluorescent lamps also receives mixtures of rare earth oxides extracted from lamp phosphors in stages by two U.S. firms that it then separates into individual rare earth oxides.

[20] Office of the Under Secretary of Defense for Acquisition, Technology, and Logistics, Report on Feasibility and Desirability of Recycling, Recovery, and Reprocessing Rare Earth Elements (September 2012). The report was prepared pursuant to a request in House Report 112-329, accompanying the Fiscal Year 2012 National Defense Authorization Act, that the Secretary "prepare a report on the feasibility and desirability of recycling, recovering, and reprocessing rare earth elements, including fluorescent lighting in Department of Defense facilities, batteries, and neodymium iron boron magnets used in weapon systems and commercial off-the shelf items such as computer hard drives."

[21] Keith R. Long et al., The Principal Rare Earth Elements Deposits of the United States—A Summary of Domestic Deposits and a Global Perspective, U.S. Geological Survey (2010); see also U.S. Environmental Protection Agency, Office of Research and Development, Rare Earth Elements: A Review of Production, Processing, Recycling, and Associated Environmental Issues (December 2012), Appendix B (listing rare earth containing mineral deposits in the United States, including currently and formerly active mine sites).

[22] Keith R. Long et al., The Principal Rare Earth Elements Deposits of the United States—A Summary of Domestic Deposits and a Global Perspective, U.S. Geological Survey (2010).

[23] 15 CFR Part 700 (Defense Priorities and Allocations System); DoD 4400.1M, Department of Defense Priorities and Allocations Manual, Feb 21, 2002.

[24] While most DoD orders receive DPAS ratings (DO and DX), a contractor will ordinarily not move a DPAS rated order (DO or DX) ahead of an unrated order unless it is necessary to meet the delivery schedule for the rated order. Note also that bilateral security-of-supply arrangements with foreign governments allow DoD to request priority delivery for contracts and orders from companies in signatory countries.

[25] The NDS process monitors and assesses risk related to over 100 individual materials.

[26] As noted elsewhere in this report, the 2013 NDS Requirements Report identified rare earth shortfalls for essential civilian demands.

[27] Annual Industrial Base Capabilities Report to Congress, submitted by the Under Secretary of Defense for Acquisition, Technology and Logistics, Office of the Deputy Assistant Secretary of Defense for Manufacturing and Industrial Base Policy, October 2013.

RELATED NOVA PUBLICATION

RARE EARTH ELEMENTS IN THE U.S. AND CHINA

Joseph A. Meyer and Daniel M. Rodrigues

ISBN: 978-1-61324-500-2 Publication Date: 2011

Rare earth elements (REE) are essential for a diverse and expanding array of high-technology applications. REE-containing magnets, metal alloys for batteries and light-weight structures, and phosphors are essential for many current and emerging alternative energy technologies, such as electric vehicles, energy-efficient lighting and wind power. REE are also critical for a number of key defense systems and other advanced materials. This book provides a nontechnical overview of domestic reserves and resources of REE and possibilities for utilizing those resources. At the present time, the U.S. obtains its REE raw materials from foreign sources, almost exclusively from China.

INDEX

D

E

F

N

O

P

Q

R

S

T

U